Maths for Adults

Studymates

Many other titles in preparation

Studymates

Helping You to Achieve

Maths for Adults

Graham Lawler

www.studymates.co.uk

First published in 2005 by Studymates Limited, PO Box 2,
Bishops Lydeard, Somerset TA4 3YE, United Kingdom.

Telephone: (01823) 432002
Fax: (01823) 430097

Typeset by PDQ Typesetting, Newcastle-under-Lyme
Printed and bound in Great Britain by The Baskerville Press Ltd.

Contents

v

Preface

2010 ... in 2010 the world is set to change. How do we know this? Quite simply, because we can 'read the numbers'. At the time of writing this is just five years away. In five years we are due for a major shock. The fact is that 77 million Americans hit 65 years of age in 2010. These are the 'baby boomers' and they are heading for the future with unrealistic demands. They expect social care and they expect their pensions to be in place. The point is that the money paid in taxes into pension pots does not actually get invested at any rate of interest. The money is used to pay members of the current generation their pensions. This means that in 2010 there will be a huge demand for payment – but who will pay? The younger generation is smaller than the baby boomer generation. This means that there will be an increased burden on each taxpayer to pay for more pensions. Is this scaremongering?

This book was written in the UK so why should Europe worry about what happens in the US? Partly because a very similar pensions system is in operation here too, but mainly because America is the dominant world power of the moment. If America suffers an economic downturn, there is no doubt that it will drag many other countries down with it. On average in each lifetime there are two recessions and a depression. We have had two recessions – in the 1970s with the oil crisis and again in the late 1980s and early 90s – but we have not had a depression. Gloomy isn't it? Frankly it *is* gloomy unless we use the next five years to protect ourselves. So how can we do that?

There are a number of steps we can take and the first is education. It is basically about understanding what is happening in the economy. That is the point of this book. Many readers are making serious mistakes and need to change the way they think about money. In this book I want to show you how, with a little understanding, you can develop a pattern of life which will set you on course to become wealthy. It involves changing your thinking – it is about understanding what the numbers are telling us and about planning your future.

To do this you need three things:

- a vision;

- the focus to work towards that vision;

- the persistence to keep going when times are bad.

Vision

Imagine where you will be three years from now. Make a detailed picture in your mind of how you are going to be a more rounded person with a greater depth of knowledge than you have ever had before. Be excited, this is the start of a new life and you *can* make it happen. In your mind, imagine the high quality home you will live in and the nice car you will drive, smell the high quality leather seats in your car and enjoy the moment. Think about the deeply warm personal relationships you will have because you are a more interesting and informed person than you have ever been before. Spend some time and develop your vision of your future – this is the first step to your new life. It is a good idea to start a private journal. I have written about the concept of this journal in more detail in other books – it really does work.

Journal	• Begin by describing your vision.

Focus

Having decided on a pathway, you need the focus to stay committed to it. The England soccer star David Beckham is totally focused on football. When he played for Manchester United he stayed on the training ground longer than most of the other players simply to practise his now famous free kicks. When you are learning mathematics *you* need to provide the focus. This means that you basically need to work on the mathematics and actively get involved. The only way to learn mathematics is to do mathematics. There is no short cut. If you are willing to focus on the job and to do the work involved, you will be successful.

Persistence

Having established your vision and focused on what you want to achieve, you must keep on going. There are many people who will tell you they simply haven't got the time. In fact we all have 24/7 – it is simply a matter of how we prioritise the tasks we need to work on in the time available. The best way forward is to make a 'time plan' for your week and stick to it. This is easier said than done and it does mean managing your time and explaining what you are doing to the rest of the family. Make sure that if you disappear into another room to do some reading, you follow up by giving some of your time to others in the family. To succeed you must persist, and the success will be well earned.

The structure of the book

The purpose of this book is to show you how to get started – how to take charge and learn the first little steps that will start you on a new life in which you can avoid the financial pitfalls that lie ahead. I hope I am wrong in my gloomy forecasts above. If I am wrong maybe I will look and feel a little foolish, but if I *am* right what will you feel like?

Don't feel bad when you make mistakes – everybody does at some time or other. I certainly have when it comes to finances and I wish I had known 30 years ago what I know now. Take the first step and educate yourself. The future belongs to the informed citizens so make sure that you are one of them.

The book is written in three parts.

Part One covers the basic mathematics that all adults should know. The old excuse that 'I could never do maths when I was at school' is now old-century thinking. To be able to deal with the demands of adult life it really is important that you understand the basics as they are explained here.

Part Two covers data handling. Many people have to handle data in their jobs and, frankly, do not understand what the numbers are telling them. Remember that the whole tenure of this book is basic mathematics so don't panic – with a little thought it will be fine.

Part Three covers money. The love of this may be the root of all evil to some, but the fact is that we live in a world where it is the medium of exchange. The capitalist system is based on the exchange of goods and services for money. Understanding and controlling the flow of money, so that more flows to you and less away, is vital. The examples used in this section are necessarily simplistic – they are not meant to be prescriptive but simply to 'set the scene' and trigger your imagination so that you can relate your own circumstances to a, possibly, different future.

I and we

In writing this book I have deliberately interchanged 'I' and 'we'. The fact is that most books are the result of a team effort and I wanted the team to gain recognition. Since I am the one who actually put the words on the page I am the one who gets to have my name on the cover, but the fact is that editors, typesetters, proofreaders and mathematics experts have all played a part in producing this book and they deserve some recognition. So it is only right that I put on record my thanks to Tony Clappison, my editor; to my colleague Jane Furlong of Soar Valley College, Leicester for her professionalism and unstinting good advice; to Susanne from Publishing Services for her support in the professional aspects of developing this book; to Marian and Pete of PDQ, the typesetters; and to Jan Boonstra from Amsterdam, who is clearly a bright student.

However the one person to whom I am indebted for her love and support is my partner and wife Judith, who once again has had the patience and stamina to endure the domestic disruption that writing a book such as this brings to a home.

I do hope this book helps you and that you will benefit from using it. Please do let me know how you are getting on, either through my publishers Studymates (graham.lawler@studymates.co.uk) or through my own website at www.mreducator.com. I look forward to hearing from you.

Graham Lawler

Part One
Basic Mathematics

1 Calculating the Future

'*the time will come when you will hear me*'
Benjamin Disraeli

One-minute overview

In the preface I wrote a gloomy prediction of the future. I genuinely hope that I am wrong but, from where I sit, I see many people moving forward in life like poor sailors who are heading for the rocks. I don't want *you* to be the one who is shipwrecked. With a certain amount of knowledge about how the economy works and a basic understanding of the numbers, you will be able to navigate to miss the rocks and have a far brighter future than you ever thought possible.

In this chapter we will look at:

- how the world has changed and is changing, economically speaking;
- how you should plan your future.

When I was a kid, growing up in Wales in the 1960s and 70s, it really was quite easy. I was told that all I needed to do was work hard, get qualifications and then I would get a 'good job' and a great salary. I would be able to live in a huge house and would have a beautiful wife. I was also told by a number of teachers that I would have to 'go to England', something that many Welsh people over the age of 40 will recognise. Well the beautiful wife bit came true and the bit about moving to England but as for the rest, well...

'*What am I pretending not to see?*'
Donald Trump

So that was the system and I did my bit. I went to college and studied hard. I gained my first degree. I then studied part-time and got two more degrees and am now working for a PhD. But in 1995 my world changed. In 1995 I was taking a break from teaching and was working for Oxford University Press in their Somerset office. I was also working for my MA with the Open University and took a course in Economics and its effect on education and society. This was a real eye-opener because it told me that everything I had based the previous twenty years of my life on was wrong. You see, part of the study involved reading the work of Robert Reich.

Robert Reich is a very astute writer and scares me witless; you see he analysed the economy and showed how it had changed since the late 1970s. I went to college in 1978, just at the time the economy started changing, and was unaware of what was happening.

Reich created a new view of the economy and how we work within it. He created what we call a 'model' – a structure that explains what is going on. Now a few words of warning – in social science models abound like a rash of freckles on a child's face. In many cases you can pick and chose the model you want to look at. But, given that concern, I want you to consider what Reich has determined and just think about it for a moment.

Reich came up with the idea that there are three types of jobs in the world, yes only three.

- *Routine producers* are people who do routine jobs and are often semi-skilled. Typical examples include creating products from kits – for example washing machines – in shift work.

- *In-person providers* are people who do jobs that currently cannot be completed by a machine. As a teacher, I fitted this role well. Another type of in-person provider is the person who serves your fries at the fast food outlet.

- *Symbolic analysts* are people who use symbols to solve problems and then apply them to the real world. This is not the same as people who have been to university. An electrician needs to understand circuit diagrams to wire up a house properly and so can justifiably claim to be a symbolic analyst. The person who works out the minimum route for a delivery van is also a symbolic analyst.

Clearly teachers, even mathematics teachers, are not symbolic analysts – they are in-service providers.

Then the bad news hit me – since 1979 routine producers and in-person providers have seen a drop in salary in real terms. Only symbolic analysts have seen an increase in their pay packets. They have seen their pay grow by about 40% in real terms. Teachers have seen a drop in their pay since the 1970s in real terms. Different commentators disagree about how much, but generally speaking in the late 1970s the average teacher earned 130% of the average non-manual wage earner's salary. By 1999 this had fallen to 99% and is still falling – no wonder there is a shortage of teachers.

> '*Continuous effort – not strength or intelligence – is the key to unlocking our potential*'
> Winston Churchill

What does this mean for us?

It means that the old ways of working are over – they don't work any more. The old ways are the traditional, middle class routes but we all know people who do not follow this pathway and are rolling in cash, while others work hard and have very little to show for it. Why is that? It is because those who work hard in the traditional way, and I was one of them, have failed to see the world change around them.

Here is an example to make you think. There is a young teacher in the South of England who got a degree after four years of study, he earns about £18,000 (about $30,000) per year. He has a nice job and a young family to raise. If he were to run a market stall, I am told he could make £50,000 a year. So, we have to face it – the traditional way of working has had its day. This is because it was the way of working in the industrial age. The industrial age is over. We are now in the *information age.*

As I sit writing this, I am listening to a DJ on a digital station playing my kind of music. A digital station is a radio station that simply couldn't have existed ten years ago. The widespread use of the silicon chip has meant that we can now have services that are more efficient and far cheaper than traditional methods would have produced.

I wince when I see trade unionists marching to save jobs for old industries. In the long run those jobs must go because we are now in the information age. To be successful in the information age we have to change how we think and how we work. We are living through a revolution – in the same way that our ancestors had to leave the fields and go to the towns to find work, so we have to leave muscle jobs behind and move towards thinking jobs.

Muscle jobs?

My dad is Irish so I enjoy Irish and British nationality. He is one of the many Irishmen who were forced to move to look for work. He moved to Wales and worked the rest of his life as a butcher. As a child I used to accompany him on his meat rounds travelling through many of the villages in North Wales. We would go, early in the morning, to the abattoir to collect the meat. He would then carry huge sides of beef into the van and we would take it back to the cutting room for preparation. This was incredibly hard work and while Dad, who is now enjoying his retirement, was not a huge man, he was certainly strong and wiry. In many ways he was earning money from his knowledge of meat but also from his strength in lifting and carrying meat.

As a student I worked on the buses during my summer holidays. I saw big men battle with old Bristol Lodekka buses that were built when power-assisted steering was a distant dream. This was another muscle job. Each man had to be a skilful driver but also had to be strong.

These are two examples of jobs that called for a combination of skill, knowledge and strength. In the North of England there is a lovely old word that is still used for strength, it is 'brawn'.

Look at what is happening to these types of jobs. We eat less meat now than in the 1960s and 70s and whenever heavy meat has to be moved then machines are, quite rightly, used. The jobs have suffered a drop in income in real terms. Look what is happening to other industries based on labour – they are being squeezed. This is because manufacturers are moving low-skilled jobs abroad. This is because the costs of labour are cheaper abroad. No-one in the UK would be able to live on £3,000 a year for but in countries like India this is a very good salary. The use of call centres in India is now well established. Production facilities for manufacturers are well established in China simply because the cost of labour is so much cheaper than in the UK, mainland Europe or the USA. These industries are now being called 'sunset' industries and are effectively being exported.

We are exporting jobs because we live in a global market. Many large companies now have multi-national status and are concerned with manufacturing quality products at a price at which they can both sell and make a profit. The only way a country like the UK can compete is to engage its workforce in jobs that 'add intellectual value'. These are jobs that a machine or low-skilled /semi-skilled person cannot do.

In the information age we need brain jobs. These are jobs in which personnel need specialist knowledge at all levels in the economy.

Brain jobs, drain jobs?

But what about the jobs we have exported? Will we run out of jobs?

The short answer is no. There are new types of job being created all the time. For instance there is now a 'human/ computer interaction specialist' job. This is a job in which the person has to use specialist knowledge in the creation of new ICT-based businesses, in particular where people have to interact with machines. The purpose is to create machines that fit people, rather than people having to fit the machine.

So when old jobs disappear, new jobs are created. The development of new technology brings new job opportunities. Think about what happened when electric street lamps replaced gas street lamps. Gas lamps had to be lit and men were employed to walk the streets and light each lamp. This job has disappeared but subsequent generations have moved into different jobs which didn't exist at that time. The argument given by some politically active trade unionists 'it's not your job, it's your son's job and his son after him' is naïve – life simply isn't like that, and frankly it never was.

What about the middle classes?

They are no better off than the traditional working classes. Traditional middle class wisdom has it that you need to work hard (I did that) to get qualifications (I did that too) and then all your problems are over. Er ... no they are not, and I can vouch for it. Traditional middle class wisdom also suggests that if you work hard and stay with the same company then you will get a good pension. Yet many companies have recently closed their pension schemes or altered them leaving employees in the lurch. What do you do when at the age of 63, with two years' work left, the company announces that your pension will be worth a third of what you had been promised? Sounds far fetched? This is exactly what happened in South Wales in 2004 and for the people concerned it was no laughing matter. The fact is that there is a growing number of people who have no choice but to keep on working past the age of 65. I know of one case where a 70 year old man is still working as a roofer despite being ravaged with arthritis!

So how should you plan your future?

The fact is that the world has been changing for the last 25 years and many people have yet to notice. The first step in planning your future is to make an investment in your own education. In particular, you must become financially literate. Later in the book we will discuss John Cummutta's 'debt to wealth' programme but it is essential that you realise that the status quo is not an option – you *must* take action now. It is not my place to make decisions for you, I am simply a writer who is trying to alert you to the facts that you need to think about.

You cannot sit back and expect the state to look after you in old age – there simply will not be enough cash to go around. In late 2004 it was announced that volunteers will be staffing some police stations. If we cannot afford to pay for police officers now, how will we be able to pay for everything in the future? We have to take control now.

You need to understand how to read the numbers, how to interpret what they are telling you, how to encourage your children to develop their own understanding of the numbers and how you, as a family, should develop the habits of the rich, the habit of investing in your own future.

2

Your Calculator

'...be open to new ideas and be careful about dismissing them too quickly'

Sir Clive Woodward

One-minute overview

In this chapter we will discuss the use of the calculator. The calculator is an essential tool on your desk but it can be a real threat to your and your children's understanding of how mathematics works. That said, everyone should be able to use a calculator so please do not ban their use in your home. This is not an overstatement – we have heard of many parents who state that they are banning the calculator. This is technophobia – it is like banning the use of a kitchen cooker because food can be cooked on an open fire. Just because it can be cooked on an open fire doesn't mean that it is recommended daily practice. Calculators save hours of time by eliminating drudgery and allows concentration on higher level tasks.

In this chapter we will look at:

■ how to use fractions on the calculator;
■ how to use memory;
■ how to use the second function keys;
■ the meaning of the term 'reciprocal' and its use if your calculator has a reciprocal button;
■ how to use the ANS button if your calculator has one.

We suggest that you spend some time working with your calculator because it is important that you become familiar with it as a machine. Using a calculator is reminiscent of driving a car. After you have been driving for a while, you automatically reach out where you know the correct buttons and switches are located. You really do need to be that familiar with your calculator.

Incidentally, make sure that your children have the correct calculator months before any examination so that there is time to get used to it. We have heard of many instances where students have said they will get the appropriate calculator just before an examination. This is not a good idea – it is like practising for your driving test in an old banger and then taking your test in a brand new car in which you are not familiar with the location of the buttons. It simply doesn't make sense.

Why use a calculator?

Quite simply, the calculator is a tool that frees you from the tedium of crunching numbers. It means that you are empowered to calculate more often and at a deeper level of understanding than would have previously been possible. Readers who were in school up to the mid 1970s will remember the agony of having to work, for example, with logarithms as a calculation aid without really understanding what was meant by the terms log or antilog, just using them to crunch numbers.

What the majority of the population determine as mathematics – namely adding and subtracting, mentally calculating percentages, fractions and decimals – is in fact only a part of mathematics. Real mathematics does involve the application of strategies and knowledge in the solution of problems. However, using a calculator frees the brain to think at a higher level. But, and this is a vitally important point, some people do become calculator dependent. We have seen people use a calculator to calculate 3×2 and that can never be acceptable practice.

You should use the calculator sensibly and for more complex calculations because this is normal practice in everyday life.

The keyboard layout

In this section we are going to look at the layout of a scientific calculator. We will not, at this stage, look at the basic four-function calculator. If you have not bought a calculator then we recommend that you buy a scientific version because of the way they perform calculations. In fact if you have children over the age of 11 then this is a far better option. We will carry out a simple investigation shortly and you will see what we mean.

Why is there writing between the keys?

If you look at the keyboard of a calculator, you will see various bits of writing between the keys. The reason for this is that each button has two jobs (functions). The alternative would be to have a machine twice the size. It would be far more expensive and not sell well.

How do I enable the second use of the button?

On your keyboard you will find a button, usually at the top of the keyboard, marked '2nd F' or 'INV' or 'SHIFT'. Pressing this button activates the second function of the button.

The main keys for basic mathematics

The fraction key

On the keypad look for a button marked $a^{b/c}$. This is the fraction key. To enter the fraction ½ you need to press '1' then '$a^{b/c}$' and then '2'. The display will look something like a 1 followed by reverse L shape and then a 2.

Similarly if you want to key in a mixed number, such as $1^1/_2$, press '1' then press '$a^{b/c}$' then press '1' again, press '$a^{b/c}$' again and then '2'. The display will look something like 1 ⌐ 1 ⌐ 2.

The $^+/_-$ key

This key interchanges the sign of the value in the display. To enter −5 on most machines you will need to press '5' and then '$^+/_-$'.

The $\sqrt{}$ key

This symbol means 'square root'. The square root of a number is the number that you must multiply by itself to generate that first number. For example, the square root of 100 is 10 because $10 \times 10 = 100$. The square root of 25 is 5 because $5 \times 5 = 25$. When you write $\sqrt{36}$ it is the mathematical equivalent of saying 'the square root of 36'. OK, so what is the square root of 36?

You will need to examine your calculator. Some calculators work in the sequence that you would expect. You press the $\sqrt{}$ button first and then you key in the number you want to find the square root of, for instance 36. So you would press '$\sqrt{}$' and then '3' then '6' and then '='. This should give the answer 6. Other calculators work on the sequence '3' then '6' and then '$\sqrt{}$'.

You should ensure that you and your children (around 11 to 13 years old) can recall the square roots of all of 1, 4, 9, 16, 25, 36, 49, 64, 81, 100, 121, 144, 169, 196 and 225. Get your children to make up a poster of square root numbers.

The x^2 button

This is the square button. When you square a number, it means you multiply it by itself. For example 10×10 is mathematically said to be '10 squared' this is written this as 10^2. To key this on your calculator press '10' and then 'x^2'. You should get the answer 100 (because $10 \times 10 = 100$). In the same way, $5^2 = 25$.

The $^1/_x$ button

This is a very useful button because it helps to work out the reciprocal of a number. The reciprocal of a number is the number that you multiply the first one by to get an answer of 1. For example, the reciprocal of 2 is ½ because when you multiply them the answer is 1. So, if you want the reciprocal of a number simply write it as a fraction with 1 in the numerator (top) and your number in the denominator (bottom) – or simply key in the number and press the $^1/_x$ button. You need to be aware that the machine will usually give you the answer as a decimal.

'*Technology can accelerate a transformation*'
Jim Collins

The memory buttons

The lettering on the memory buttons will depend on the type of calculator you have. Some machines have a large number of memory banks. In the *Mr Educator* office we have a now rather old programmable calculator that has 64 memory banks. You will usually find the memory buttons marked M+ and MR.

The M+ button can be used to input information into the memory bank. Press '6' and then 'M+' and you will add the number on the display to what is stored in the memory.

Press 'AC' to clear the display. The MR button is the memory recall button. If you press it now 6 should reappear in the display.

You should also find a button marked $M-$ (or it may be the second function on the M+ button). Pressing this button subtracts the number in the display from what is stored in the memory bank. Remember, if it works as a second function button you will need to press the '2nd F' or 'INV' or 'SHIFT' button first.

The x^y button

Pressing this button calculates any number (x) to a power (y). For example 5^3 means '5 to the power of 3' or $5 \times 5 \times 5 = 125$. If you press '5' then 'x^y' then '3' and then '=', the display will give the answer 125.

You may find that your calculator has y^x instead of x^y, or it may be a second function on your calculator keypad.

The ANS button

Some calculators have an ANS button. These calculators store previous answers that you have worked out. Suppose you are working through a set of steps in a calculation and you forget a previous answer in the chain, pressing the ANS button retrieves the previous answer.

Different types of calculators

Here we are going to investigate different types of calculator. You need a scientific and a basic four-function calculator. On both calculators, work out the following:

$$1 + 2 \times 3 =$$

What do you notice? Can you explain what is happening and why it is happening? Does this always happen?

Try other calculations like this – does the same thing happen? Try this with your children and ask them what they think is happening here.

It is important to understand the mathematics that is going on here. There is an *order* of operations in mathematics – by this we mean there is a correct sequence of doing things. The two calculators have come to different answers because they work in different ways. They cannot both be right and they are not. They have different logic systems.

The scientific calculator is the one that gives the 'correct' answer. The easiest way to remember the mathematical order of operations is that $\times$ and $\div$ come before $+$ and $-$. So with $1 + 2 \times 3$ we first of all calculate the 2×3 to get 6, only then do we add the 1. A clearer way of writing this is to use brackets, as in $1 + (2 \times 3)$.

The basic four-function calculator works on a different logic system. It does the calculation in the order that it is written. So it works out $1 + 2$ to get 3 and then it does 3×3 to get the answer 9. Again, if this is the way this particular calculation is meant to be carried out then it should be written with brackets, as in $(1 + 2) \times 3$. This indicates that the addition has to be calculated first.

More on brackets

By using brackets you 'tell' the calculator about the order in which *you* want the calculation to be carried out, otherwise the calculator will work out the answer by doing multiplication and division before addition and subtraction. Try working out $(1 + 2) \times (4 + 5)$ in one go on your calculator. You should find that the answer is 27.

What about double brackets?

Let's say you have to work out 3×4 first, then add 5 to this answer, and then multiply it all by 10. How would you write this down? You need to use double brackets. We usually use square brackets [ ] followed by round brackets () to make things look clearer. So the calculation described above is written as

$$[(3 \times 4) + 5] \times 10$$

To check that you have written the brackets correctly, make sure you have an even number of brackets. Here the round bracket in front of the 3 matches with the round bracket after the 4, and the two square brackets link.

Note that in the order of precedence, brackets come before multiplication and division. In effect this means that you work from the inside set of brackets towards the outside set.

What is the DRG button?

Most basic mathematics is done with the calculator working in 'degree mode'. Your keypad will have some kind of degrees, radians or gradians button, it may be a second function, or a mode function. Read your calculator booklet carefully to see how it works. You will almost certainly have to work in degree mode, and note that if you are in another mode you will probably get wrong answers.

Now the next task is to deal with some basic number operations.

3

Dealing with Fractions

'Thinking is the hardest work there is, which is the probable reason why so few people engage in it'
Henry Ford

One-minute overview

Calculating quantities is a basic skill need to improve both your mathematical knowledge and your financial literacy. You can usually buy a large quantity of a product at a cheaper unit rate. For instance, large cans of food are often cheaper pro rata than smaller cans. The large cans may contain twice the contents but they are not twice the price, so the ability to calculate a unit cost is important even at this mundane level.

In this chapter we intend to take you back to those heady days when you were at school and were asked to add fractions but never 'got it'. If you did 'get it' then congratulate yourself but the fact is that this is something most adults cannot do. With a bit of clear explanation it will become clear and this really is an important skill for dealing with quantities.

In this chapter we will look at:

- what a fraction is;
- the equivalence of fractions;
- how to add, subtract, divide and multiply fractions;
- how to convert between improper fractions and mixed numbers.

What is a fraction?

The word fraction means 'part of a whole'. If a whole is broken into parts, the parts are fractions. As you know, $^1/_2$

is a fraction but what you may not realise is that $^1/_2$ means 1 part out of 2.

Look at this diagram.

The large rectangle is split into four smaller parts. In Europe these are called quarters, in the USA they are called fourths. So clearly $^4/_4$ is a whole 1. In Europe this is read as four quarters, in the USA this is read as four fourths – either way, it is clearly equivalent to a whole 1.

Naming the parts of fractions

The top part of a fraction is called the numerator and the bottom part the denominator.

$$\frac{\text{numerator}}{\text{denominator}}$$

A fraction is a fraction, of course, but there could be some confusion with some numbers. These are the kinds of fractions you might come across:

- a vulgar fraction is a simple fraction, like $^3/_4$, with the numerator smaller than the denominator;

- an improper fraction is a top-heavy fraction, like $^{15}/_3$, with the numerator larger than the denominator;

- a mixed number is one like $2^1/_4$.

What is the connection between the denominator and the size of the fraction?

Some people find it quite bewildering that $^1/_2$ is bigger than $^1/_5$ when 5 is obviously bigger than 2.

It seems a bit of a contradiction to some until it is considered more closely. Look at these diagrams and imagine they represent two cakes of the same size – one has been cut into two slices and the other into five.

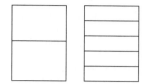

It is quite clear from these diagrams that 1 part out of 2 must be larger than 1 part out of 5 (so long as we are talking about the same whole).

What happens if the numerator and denominator are multiplied by the same number?

Look at this fraction, $^1/_2$. Multiply both the numerator and the denominator by 2. What happens?

$$\frac{1 \times 2}{2 \times 2} = 2/4$$

Now we know that one half is indeed the same as two quarters (or two fourths).

Equivalent fractions

Using the cake analogy again, we can cut the cake so that, say, two slices of one cake are the same as four slices of the other cake.

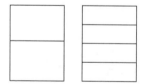

This diagram shows the two cakes cut into halves and quarters. From the diagram it is easy to see that $^1/_2$ is the same as $^2/_4$.

But drawing diagrams like is a real pain, and frankly is unrealistic when we want to work out equivalent fractions.

Families of fractions

Families that are 'equivalent' are said to belong to the same 'family'. In other words, fractions in the same 'family' are all equal.

An easy to see what we are saying is to look at the family of fractions for $^1/_2$.

We have already shown that $^1/_2 = {}^2/_4$. But we can carry on this sequence

$$^1/_2 = {}^2/_4 = {}^3/_6 = {}^4/_8 = ...$$

Compare the numerator and denominator of each fraction and write down what you notice.

You should have noticed that the top number of the fraction is half of the bottom number of the fraction. In other words, whenever a numerator is half of denominator the fraction is equal to $^1/_2$.

The same can be said of the family of thirds. Convince yourself by starting with $^1/_3$ and then multiplying the numerator and denominator by the same number. This will generate the family of thirds.

How to do calculations involving fractions

Fractions can be multiplied, subtracted, added and divided. But how? They are not whole numbers. How can $^1/_2$ and $^1/_3$ be multiplied? How can you work out $^1/_2 \div {}^1/_3$? How on earth do you add $^1/_2$ and $^1/_3$? And what about $^1/_2 - {}^1/_3$?

Just like other mathematical problems – by using a set of simple rules.

How to multiply fractions

This is probably the easiest of the fractions sums because you just multiply the numerators together and then the denominators. So $^1/_2 \times {}^1/_3$ becomes $\frac{1 \times 1}{2 \times 3}$ which is $^1/_6$.

There is just one other thing to note. If the final answer can be made simpler (with a smaller denominator) then it should be. So $^3/_4 \times {}^2/_3$ becomes $\frac{3 \times 2}{4 \times 3}$ which is $^6/_{12}$. But $^6/_{12}$ is the same as $^1/_2$ and this is a 'simpler' fraction and so is the 'proper' answer, i.e.

$$\frac{3 \times 2}{4 \times 3} = \frac{6}{12}$$

$$^6/_{12} = {}^1/_2$$

How to divide fractions

Dividing fractions worries many adults because they were taught to do it routinely without any understanding of the mathematics involved. Let's think about a straightforward calculation:

$$1 \div {}^1/_2$$

Think about what this statement means. It is asking you a question and that is 'How many halves are there in one whole?' The answer must be 2.

Now apply the same reasoning to this calculation:

$$^1/_2 \div {}^1/_4$$

This statement is asking the question 'How many quarters are there in $^1/_2$?' So it seems reasonable to expect the answer to be a whole number.

Journal	• Think about this for a moment and make an entry about the basic idea involved.

To divide a fraction by another fraction, you simply invert the last fraction (turn it upside down) and then multiply. So here it means

$$^1/_2 \div {}^1/_4 = {}^1/_2 \times {}^4/_1$$

$$= {}^{1 \times 4}/_{2 \times 1} = {}^4/_2 = 2$$

In other words there are two quarters in $^1/_2$. Or

$$^2/_4 = {}^1/_2$$

But why does this inverting method work? Let's look at the underlying mathematics. If you were asked to divide €1 by 2 most people would quickly say 50 cents. There is another way of working this out, instead of dividing by 2 you could invert it and multiply by $^1/_2$. In other words $^1/_2 \times 100$ cents also gives the answer 50 cents.

2 and $^1/_2$ have a special relationship – they are called reciprocals of each other. Remember, reciprocals are numbers which give an answer of 1 when multiplied together.

Dividing by a particular number has the same effect as multiplying by the reciprocal of that number. Try some for yourself. Work out $300 \div 6$ and then work out $300 \times {}^1/_6$. What do you notice? You should see that they both give the answer 50.

Journal	• Think about this for a moment and describe what you have done here.

How to add fractions

To be able to add (or subtract) fractions you must make sure that they have the same denominator. $^1/_2$ and $^1/_3$ don't, so they must be converted. Using the family of fractions method, show that they can be added – don't worry, we'll wait.

Journal	• Record your working and then you can compare what you have done with our answer below.

The first step is to work out an equivalent for $^1/_2$, then work out a similar equivalent (with the same denominator) for $^1/_3$. Then and only then will we be in a position to add the fractions.

Using the family of fractions idea we get $^1/_2 = {}^3/_6$. Similarly $^1/_3 = {}^2/_6$

Now you can see that $^1/_2 + {}^1/_3$ is the same as $^3/_6 + {}^2/_6$, which gives us $^5/_6$

Let's try another example – calculate the value of $^1/_4 + {}^1/_5$ The overall method is the same – first work out the appropriate equivalent for each fraction and then add them.

It helps to work out the *lowest common denominator* (remember that from school days – or was it daze?). This is the lowest number that both 4 and 5 will go into, in this case 20.

Now $^1/_4 = {}^5/_{20}$ and $^1/_5 = {}^4/_{20}$ and adding these together gives $^9/_{20}$.

> *Health warning*
> **It is not always a case of multiplying the denominators to find out which fractions to use for equivalence. For instance, suppose you wanted to add $^1/_4$ and $^1/_2$. Multiplying these denominators gives 8 but the lowest common denominator in this case is 4 – the lowest number that both 2 and 4 go into.**

How to subtract fractions

We can use and adapt the 'families' technique that worked for adding fractions to subtract one fraction from another. For instance, to calculate the value of $^3/_4 - {}^1/_3$ you first work out an equivalent of $^3/_4$ and then do similarly for $^1/_3$.

The lowest common denominator is 12 and so $^3/_4 - {}^1/_3$ becomes the equivalent $^9/_{12} - {}^4/_{12}$ and the answer is $^5/_{12}$

Improper fractions

The fractions we have dealt with so far are called vulgar (or simple) fractions. An improper fraction, for instance $^3/_2$, is commonly called a 'top heavy' fraction. But what does this mean? This diagram shows three halves arranged in a certain way.

Looked at this way, you can see that $^3/_2$ obviously is the same as $1^1/_2$. Numbers such as $1^1/_2$ are called mixed numbers.

Remember that a fraction over itself is equal to 1, so all top heavy fractions must have a value greater than 1.

Changing mixed numbers to improper fractions

Think of a mixed number in terms of its diagram, for example think of $2^1/_2$ as

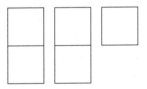

You can see now that $2^1/_2$ is the same as $^5/_2$.

Let's try another – but no diagram this time. Convert $5^1/_4$ to an improper fraction.

Now let's think this through logically together. We know that '5' represents 5 whole ones. If each of these is cut into quarters there will be 20 quarters (5×4). All we need to do then is add the other quarter – so this means that $5^1/_4$ must be the same as $^{21}/_4$.

Hardly awesome is it? Just a simple technique that, with a bit of practice, you can use to solve these types of problems.

Adding and subtracting mixed numbers

This is more straightforward than it first appears. Look at $10^1/_2 - 3^3/_8$ for instance.

This calculation can be broken down into three steps:

- work out the whole numbers first, namely $10 - 3 = 7$;

- use the equivalence idea to work out the fraction bit, namely $^1/_2 - ^3/_8$ is $^4/_8 - ^3/_8$ which is $^1/_8$;

- add the answers to the first two parts to get $7^1/_8$.

Fractions of quantities

This is the most important section in this chapter because this is probably the most common area where you will need to use fractions.

For example, you might have to calculate $^3/_4$ of 100. This kind of calculation is quite straightforward and many can be done quickly in your head – but bear with us because the *method* is important.

- Mentally work out $100 \div 4$. This gives an answer of 25 and is, in effect, one quarter of 100.

- Multiply this by 3 to get 75, which gives three quarters of 100.

This is a useful technique – in short, it is find one lot of the fraction and then multiply by the required fraction numerator.

Final comments

This chapter is intended to be a quick reminder of the basic numeracy work you will have dealt with at school. Use it as a resource to dip into for reminders – and enjoy using the simple techniques described.

The point about these techniques is that they are invaluable when dealing in business. In particular, it is important to feel comfortable when converting between fractions, decimals and percentages. You will find that as soon as you are clear in your own mind that, for instance, $^1/_5$ is 0·2 and is also 20%, the greater an advantage you will have. The reason for this is simple – this level of knowledge is beyond many people and that means you are ahead of the pack.

Journal	• Write down a sentence that describes one aspect of your new personality, your new found power.

'If you think you can or think you can't, you're right'
Henry Ford

4

Dealing with Decimals

'Do what you can, with what you have, where you are'
Theodore Roosevelt

One-minute overview

Do you worry about dealing with decimals? If so don't despair, you are not alone. One of the commonest causes of concern over decimals is that people do not *understand* what they are doing and have no real *understanding* of what decimals mean.

In this chapter we will look at:

- the meaning of the different place values in decimals;
- how to change decimals to vulgar fractions;
- how to round off to decimal places;
- how to work with decimals when they are representing money;
- how to use a calculator to deal with money.

The meaning of place value

Have you ever heard someone read 0·25 as 'nought point twenty-five'? That person does not understand the concept of place value. Place value refers to the value that a figure is worth because of its position in a number.

For instance, if you have the number 654 how much is the 5 worth? In this case it is 5 tens, in other words 50.

But how much is the 5 worth in 645? We are still looking at a 5 but here it is one place to the right so it is worth ten times less than its value in 654. This means it is worth 5 units, in other words just 5.

So the location of a figure in a number is very significant. What about figures after the decimal point? The first thing to understand is that figures to the right of the decimal point are expressed individually. For instance 0·56 is said as 'nought point five six' and not 'nought point fifty-six'.

Think about how you learnt about whole numbers as a child. The chances are you were introduced to units, tens and hundreds in their respective columns. In the same way, the figures in decimal numbers are placed in columns which signify their value. The decimal part of the number is made up of the decimal point and the figures written to the right of the decimal point.

There are columns in the decimal part too. Reading from left to right away from the decimal point the values are tenths, hundredths and thousandths.

Notice how the decimal point is to the left of the tenths column. It is useful to think of the decimal point as being static with the figures moving around it.

Journal	• Write down some decimal numbers with different numbers of 'columns'. • Draw the number columns in one colour and use another colour to make notes explaining what they mean.

The tenths column

This is the decimal equivalent to dividing by 10: in other words, $0·1 = {}^{1}/_{10}$. Any figure in the tenths column has the same value as that figure written as a fraction over 10.

For instance, $0·3 = {}^{3}/_{10}$, $0·5 = {}^{5}/_{10}$ and so on.

Notice that when the figure in the units column is zero, we put a zero in that column. It is mathematically correct to write a decimal number without the zero, for instance 0·25 has the same value as ·25 – but you are strongly advised to use the zero and avoid the alternative (lazy) way of writing decimal numbers.

You need to be aware that 0·3, 0.30 and 0·300000000 all
have the 3 in the same place and, therefore, all have the
same value. Just remember is that you can have as many
zeros after the 3 as you like – since none of them move the
3 and so its value is not changed.

People do read numbers wrongly and if there is no zero in
the units column then errors can be made. There is
reported to be a case in the 1960s in which a nurse misread
a decimal number and wrongly injected a baby. The story
goes that she injected ten times too much medicine and as a
result the baby died. We do not know how accurate this
story is but it certainly provides a compelling reason for
getting it right!

The hundredths column

The hundredths column is the second column to the right
of the decimal point. Any figure in the hundredths column
has the same value as that figure written as a fraction over
100. For instance, $0·07 = {}^{7}/_{100}$. So the value of figures in
this column is small.

The thousandths column

Following on from the above, the value of figures in the
thousandths column is even smaller. $0·002 = {}^{2}/_{1000}$,
$0·009 = {}^{9}/_{1000}$.

How to change decimal numbers to vulgar fractions

Taking 0·3 as an example. You will notice that the figure 3
lies in the tenths column so it must be worth ${}^{3}/_{10}$

Taking 0·45 as an example. You will notice that the figure 4
lies in the tenths column so it must be worth ${}^{4}/_{10}$. The
figure 5 lies in the hundredths column so it must be worth
${}^{5}/_{100}$. These two fractions must be added:

$$ {}^{4}/_{10} + {}^{5}/_{100} = {}^{40}/_{100} + {}^{5}/_{100} = {}^{45}/_{100} $$

So, the decimal number 0·45 is $^{45}/_{100}$ expressed as a vulgar fraction. But it would be better to write it with the smallest denominator possible – that is done by 'cancelling'.

Journal	• Explain about changing decimals numbers to fractions.

Cancelling fractions

This was always a real trial for many people at school, but no more – technology has roared to the rescue. Assuming that you are using a scientific calculator, you will find a button marked $a^{b/c}$.

This is the fraction button and it will change a decimal number to a vulgar fraction – and also give the fraction in its lowest terms. Let's say that you want to convert 0·5 to a vulgar fraction – from the work above, you know that 0·5 is $^{5}/_{10}$. If you key in '5', '$a^{b/c}$' '10' on your calculator it will express the fraction in its lowest terms, $^{1}/_{2}$.

If you don't have a calculator then it is quite a simple procedure to work this out in your head. Look at $^{5}/_{10}$ again. How do we work its value out in its lowest terms?

You need to find the largest factor that will go into both the numerator and the denominator of the fraction. Here, the largest number that goes into both the top and bottom numbers is 5. So divide both by 5 and that will give the answer, $^{1}/_{2}$.

$$^{5}/_{10} = \frac{5 \div 5}{10 \div 5} = {}^{1}/_{2}$$

You can use the same approach on any decimal fraction. Try it out now on $^{5}/_{100}$.

Journal	• Work it out and make notes on the *method* involved.

Health warning

If you are a parent you need to realise that your children must be able to use this method to convert fractions – it is not enough for them to be able to use the fraction button on the calculator. There are a number of examinations where children are not allowed to use a calculator.

Changing vulgar fractions to decimal numbers

This also is very straightforward – but you still need to understand what you are doing. For instance, what does a fraction like $^1/_2$ really mean? It means 1 whole has been divided into 2. To change a fraction into a decimal number you divide the numerator by the denominator – in this case divide 1 by 2 to get 0·5

It might help to write the 1 as 1·0 because it does not affect the place value and somehow, it is easier to end up with a decimal number required.

Note that the decimal number we have calculated starts with '0 point something'. That was to be expected because $^1/_2$ is less than 1, so correspondingly its decimal equivalent must also be less than one.

Journal	
	• There is a good thinking exercise that you can use at times like this – it is *convince yourself and then convince another person*.
	• We want you to become very familiar with this strategy because it really is helpful in developing your own understanding of mathematics. If you are a parent, then we urge you to use this thinking strategy with your children. Think of it as a mental tool. A mathematician needs to be able to dip into a mental toolbox and pull out a tool in the same way that a tradesman will dip into a toolbox and collect a practical tool. The skill is *knowing* which tool is the most appropriate to solve the problem at hand.

- By going through the process of convincing yourself, you have to think through what you have done and how you have done it. The chances are that if you have made an error during this stage, then your new thinking will clear it up. However, if you have made an error and not noticed it, then going through the process of convincing somebody else should point out that error.

Now try this one – change $^1/_4$ into a decimal. Did you convince yourself? Did you convince another person?

You should have found that the answer is 0·25, and you can confirm this on your calculator.

How to round decimal numbers

In mathematics, rounding means approximating – or giving an answer that is pretty close to the actual answer. You do need to understand that a rounded answer is not exactly the same as the actual answer, it is a close approximation. You may find that your children struggle with this idea so be patient with them, it is an important life skill.

Journal

- ...OK we keep asking you to make notes and it may be starting to irritate but please do stick with us. The fact is that we are encouraging you to interact with the text because this will lead to you understanding more. The act of writing something down really does increase level of understanding of that idea and helps the learning process.

Just pause for a moment and think about the number of times that you approximate or round up. For instance, we say that people are in their thirties or forties – we are approximating their age. When we say we were travelling at about 30 mph or that it takes about three hours to get from where I live to the centre of Chester we are rounding. In many aspects of daily life we have to approximate and we intuitively know that, while it is an approximation and close to the actual answer, it is not *equal* to the actual answer and that is a very important point.

In mathematics there is a useful symbol, $\approx$, that means 'approximately equal'. For instance, 43·4 $\approx$ 43 reads as '43·4 is approximately equal to 43'. But what are the rules for rounding numbers?

Suppose, for instance, we want to round 1·8 to the nearest whole number. The best way to think of 1·8 is as an interval on a number line.

1_____1·8_____2

Clearly, 1·8 is closer to 2, so it must round to 2.

The 'five rule' helps to avoid drawing diagrams every time you want to round a number:

- if a number ends in 5 or more, then round up – this is why 1·8 rounds to 2;

- if a number ends in less than 5, then round down.

However, it is worth using the interval idea with your children to make sure that they understand why we round up or down. It is also worth pointing out that prices do not usually round according to mathematics but with increased profit margins in mind!

Here is another example to make sure that you are clear about what you are doing – round 1·24 to the nearest whole number.

Again, think in terms of the interval on the number line.

1·2_____1·24_____1·3

It is clear that 1·24 is closer to 1·2 and so it rounds to 1·2 Apply the 'five rule' and confirm this for yourself.

Health warning

With your own children, we really do recommend that you make the effort to draw the number line. Make sure you draw it on graph paper, or paper with some sort of grid printed on it. This allows them to see and count the decimal places between the two ends of the interval. This will indicate how much they understand. If they fail to recognise that there are 10 parts between 1·2 and 1·3, for example, on the number line above then we suggest that you do not make an issue out of it. Showing exasperation, or even mild surprise, can inadvertently damage self-esteem. If they do not understand, then the easiest way forward is to investigate.

Ask them to make a judgement/guess about how many decimal places there are between two numbers, say 1 and 2. Then draw number line between 1 and 2 and label all the tenths:

1 1·1 1·2 1·3 1·4 1·5 1·6 1·7 1·8 1·9 2

Remember to count the spaces *between* the numbers and not the numbers themselves.

Repeat this with larger and larger numbers. Then ask them to make judgements by asking 'What is the same about all these gaps between numbers?'

This method of teaching gives children a real experience of having to take an active part and then reflecting on that activity. This, in our judgement, is a most effective method of teaching. But it does rely on you and your children having a mutually confident relationship – one in which undermining of self-esteem will not occur.

How to deal with money

Money, in most countries, uses the decimal system. The word 'decimal' actually means numbered in tens or proceeded by tens. The current UK system of 100 pence = £1 is a decimal system, as is the system used in most of Europe, namely €1 = 100 cents.

Avoidable mistake 1

Many people write UK currency using the £ sign first and ending with the 'p', for instance £299p. This is wrong because it is 299 pounds. If you want to write a sum like this in pence, just remember that there are 100 pence in £1 and write it as 299p. Assuming it is £2.99 and not £299.

Avoidable mistake 2

When you are dealing with money, you need to ensure that you always work to two decimal places. Why? Because the currency always needs 100 small units to make up the whole – for instance, $1 = 100 cents, €1 = 100 cents, £1 = 100 pence.

This is extremely important if you are using a calculator. To see what we mean, use your calculator to add £3·40 and £4·20, your calculator may show 7·6, but what exactly does this mean?

Think about what you know now you have worked through the decimal section earlier. You know that the 6 is in the tenths column, so it is worth 60 pence. So the answer reads as 'seven pounds sixty pence' and not 'seven pounds six pence'. If your children have problems with this, we do advise patience.

Journal	• Summarise what we have discussed in this chapter **and** write down how you feel about it. Remember your journal is private and only for you.

Decimals are an essential part of adult life – so are percentages and that is the topic of the next chapter.

5

Dealing with Percentages

'In the 21st century you will earn what you learn; if you
don't learn much, don't expect to earn much'
Graham Lawler (aka *Mr Educator*)

One-minute overview

When dealing with finance it is essential to understand
percentages. Percentages, like fractions and decimals,
cause many people anxiety. They feel that they should be
able to work with percentages and use them in everyday
life, but in truth find them quite difficult. If this applies to
you, fear no more.

In this chapter we will look at:

- the meaning of the word percentage;
- how to work out percentages of quantities;
- how to find one quantity as a percentage of another;
- how to increase or decrease an amount by a
  percentage;
- issues concerning VAT/goods tax and inflation;
- the meaning of ratios and proportion.

What does 'per cent' mean?

The word 'cent' means 100. It is a unit of currency in the
US and now in most of Europe. In the US the currency is
dollars and cents and in most of Europe it is euros and
cents.

100% of something means the whole thing. If I have one
glove I have 100% of a quantity, if I have two gloves I still
have 100% (of a different quantity) – so 100% means the
whole quantity, whatever that quantity happens to be.
Gardeners often use car trailers to carry away garden

rubbish. If my car trailer is full it is 100% full. My neighbour has a larger trailer than mine but when his trailer is full it is still 100% full – as we said before, 100% means the whole thing.

It follows from this that when my car trailer is $^1/_2$ full it is 50% full. My neighbour's car trailer is also 50% full when it's only $^1/_2$ full – even though his trailer contains more rubbish than mine (he has a bigger garden).

How to work out percentages of quantities

A percentage is a fraction with a denominator of 100, so percentages can easily be changed into vulgar fractions. For instance, 50% = $^{50}/_{100}$, 70% is $^{70}/_{100}$ and so on.

The type of problem we look at first is where you hear a statistic and want to work out how much that statistic is worth.

Example 1: What is 5% of €300?
Use your calculator to work out $5 \div 100 \times 300$. You should get 15, so the answer is €15.

Think of percentages as pennies in the pound (or cents in the euro/dollar). So in this example it is 5 cents in the euro (or 5p in the pound) for every one of the 300 euros – so it is 300 lots of 5 cents, and that is €15.

You need to remember that when you are doing money calculations on a calculator, the machine may not automatically show the final 0, so something like 56 will mean £560 (or €560 or $560)

Example 2: Work out 12% of £46.
Think of 12% as 12p in the pound, and there are 46 individual pounds so the answer is $12 \times 46 = 552$, which is £5·52

The calculation is exactly the same for euros or dollars – it's just the currency unit that is appropriate to a different country. We mention different currencies because in modern business you often find yourself working between currencies and you need to be comfortable with the idea that the mathematics is the same regardless. You would be amazed at how many people can do a calculation in one currency but faced with a different currency they go to pieces.

Example 3: Simon is a publishing executive. He was given a budget of £80 000 to develop new books last year. This year his budget has been increased by 2%.

Many people will treat this as a two-stage calculation yet it can be done in one step. £80 000 is equivalent to 100%. So if his budget is increased by 2% then the new percentage is 102%. Now we need to find 102% of last year's budget to get the size of this year's budget. On your calculator work out $1.02 \times 80\ 000$ and this will give you his new budget.

Journal	• Explain what you have done here. Why did we use 1·02 and not 1·2?

The reason we use 1·02 is that 1·2 would be 120%. In other words, it would have meant a 20% rise in the budget and not a 2% rise.

Therefore, Simon's new budget is £81 600 – and he does need to spend it more wisely than he has in the past. (Every Simon in publishing will now be worried!)

Expressing one quantity as a percentage of another

This is a big mouthful and you need to be clear what it means. You need to be able to work out the answer to questions such as what is 40 as a percentage of 160? These are everyday questions since you can come across them in the workplace or the retail park. For instance, what if you had to calculate the numbers of items left in your firm's stores before you order new items, or if you need to calculate the number of days that employees have attended work or attended training?

Example 1
To work out 40 as a percentage of 160, we first of all need to write it as a fraction:

$$^{40}/_{160}$$

The next step is to multiply this by 100. Work this out on your calculator and you should find that the answer is 25%.

Example 2
Megalaw Inc. sell widgets. At the start of the week they have 2000 widgets. On the following Saturday they had 300 left. What percentage of widgets was sold in the week?

If they had 300 left, they must have sold 1700. So the calculation is $^{1700}/_{2000} \times 100 = 85$. So they sold 85% of their stock during the week.

Journal	• Describe how we did this. Write out the example on one side of the page and make notes in red on the other side of the page. It really is worth the effort because writing it down makes it real for you.

Example 3
Jo works in a factory. In a 30-day period she had 6 days allocated as rest days and took 4 days off for illness. In this period there were also 4 Sundays on which she did not work. What percentage of the 30 days did she actually work?

It is useful to work through some calculations in steps. The first is to work out the number of days that she actually did work: 30 days total - 6 rest days - 4 sick days - 4 Sundays = 16 work days. The next step is to work out 16 as a percentage of 30.

$$^{16}/_{30} \times 100 = 53.33$$

So Jo actually worked for 53·33% of the 30 days.

However this could be a misleading statistic because it suggests that she has been absent for almost half of the time she should have been at work. Let's look at these figures a little more closely. The whole period was 30 days but Jo was not expected to work every day. 6 days were rest days so the working period is reduced to 24 days. Of those 24 days 4 were Sundays further reducing the potential working time to 20 days out of which she was absent for 4 days for sickness. This means that she actually worked for 16 out of a possible 24 days.

$$^{16}/_{20} \times 100 = 80$$

This is a clear and substantial improvement on the first set of figures and shows Jo's true attendance. She actually attended for 80% of the time that she was required to attend and is not as work shy as the first set of figures leads us to believe.

This is a good example showing how we can either be misled or mislead with figures.

Changing amounts by a percentage

You come across this frequently – you've been awarded a 23% wage rise, what will your new wage be? That television you've been fancying has now got 15% off, how much will it cost now?

Increasing by a percentage

Sometimes you will find that a quantity has to be increased by a certain percentage.

Example 1

A wagon driver delivers 400 bags of compost to a garden centre. The garden centre manager asks for this to be increased by 40%. How many bags will the next delivery contain?

We can think of the first delivery as being 100%, in other words the delivery of 400 bags was 100%. If the delivery is to be increased by 40% we need to raise it to 140% of the original order. So we need to work out 140% of 400 (so expect an answer bigger than 400):

$$^{140}/_{100} \times 400 = 560$$

In other words, to increase the delivery by 40% the new delivery must contain 560 bags, or another 160 bags, of compost.

Example 2

What if the delivery had been increased by 55%? We need to add 55% onto the original delivery of 400 bags:

$$^{155}/_{100} \times 400 = 620$$

In this case the new delivery would be 620 bags, an increase of 220 bags on the original.

This type of calculation is quite common in the working world.

Decreasing by a percentage

This is the opposite of what we have just been doing. Imagine that the garden centre manager asked the driver to reduce the next delivery bags of compost by 40%. This means the next delivery must be 60% of the original. So now we need to find 60% of 400:

$$^{60}/_{100} \times 400 = 180$$

So the new delivery would be 180 bags.

Overview of these methods

If you are increasing by a percentage, simply add the increase on to 100% and then work out that percentage of the original amount. For instance, if the increase is 45% then work out 145% of the original.

If you are decreasing an amount by a percentage, simply subtract that percentage from 100% and then work out that percentage of the original amount. For instance if the decrease is 35% then work out 65% of the original.

VAT

VAT stands for value added tax. It is added on to the price of certain goods as a percentage of the sale price. In other words, if you buy something priced at £10 including VAT then not all of that money actually goes to the seller because some tax is being collected on behalf of the government.

At the time of writing, a common rate of VAT in the UK is 17·5% for many goods. So if an article costs £10, 17·5% of that is paid to the government. Other countries charge VAT, but it may be called something else (sales tax for example), and while the actual rates may vary, the mathematics remain the same.

Certain goods are exempt from VAT, children's clothes. There is also a class of goods, meat for example, which are not exempt but on which no VAT is payable – this is because they are rated at zero (0%). This means there is no VAT payable on your weekly joint but the Chancellor could introduce it very easily at any time, although it would be politically damaging to do so.

Working with VAT

If a business has a customer who is actually another business and it is VAT registered then the customer can claim the VAT back. There are certain levels of turnover at which all businesses in the UK must register for VAT. If you exceed this level and do not register you can be fined. For more details on VAT registration contact your local Customs and Excise Office.

Some retail places, especially bulk stores, display their goods priced exclusive of VAT. This means that VAT is not included in the price shown and will be added on at the checkout. Let's look at some examples.

Example 1
A shirt costs £10·00 exclusive of VAT, so that means that VAT will be added on at the checkout. We need to calculate 11·75% of the price in order to find out the price to the customer. Do this now on your calculator – you should find it is £11·75.

Example 2
What is the final selling price of a pack of printing paper costing £5·99 exclusive of tax? £5·99 is 100% of the final selling price. We have to add on 17·5% and in one step this means we need to work out 117·5% of £5·99:

$$^{117.5}/_{100} \times 5·99 = 7·3825$$

To the nearest penny this is £7·04.

How to calculate the actual price when VAT is included

In most shops the price shown includes VAT. It can sometimes be useful to unravel the retail cost of a product

and see how much the product itself costs and how much
tax has to be paid to the government in the form of VAT.

Example 1
I buy a statue for my garden costing £80·00 inclusive of
VAT. How much was the VAT?

Here the £80·00 represents the total cost of the garden
statue itself and the tax, so it must represent 117·5% of the
actual cost. Mathematically we can write this as:

$$£80·00 = 117·5\%$$

Notice we have not used the equals sign – these two
quantities are equivalent but, strictly speaking, in
mathematical terms they are not equal.

If we know what 117·5% is worth, then we can find out
what 1% is worth and then multiply up to 100%:

$$1\% = {}^{£80}/_{117.5} = £0·6808510638$$

So 1% is worth just over 68 pence. This means that 100%
must be

$$100 × £0·6808510638 = £68·08510638$$

The statue costs £68·09 to the nearest penny. The tax paid
was £80·00 − £68·09, or £11·91.

Example 2
Imagine that you are having a swimming pool put in your
garden at the cost of £15 000 inclusive of VAT. What is the
cost of the pool without the VAT?

As in the previous example, we set up an equation where
the total cost of the pool inclusive of tax is equivalent to
117·5%:

$$117·5\% = 15\ 000$$

Now divide both sides by 117·5 to find out what 1% is worth:

$$1\% = \frac{15000}{117.5} = 127.6595745$$

So 100% £12765·96 to the nearest penny. The remaining £2234·04 is tax paid to the government in the form of VAT.

Look at the quick foolproof method of calculating VAT in the *Fast Fax* section at the end of the book.

What is inflation?

Inflation is the rise in the price of goods and services throughout the economy. In simple terms this means that goods and services are more expensive than a year ago. Economists differ in explaining the causes of inflation according to their political leanings. The main point is that inflation means that prices are going up. Even when you hear that 'inflation is falling' generally it does not mean that prices are coming down. It means that prices are still going up but not as fast. In other words it is the rate of increase that has fallen, and not the prices themselves.

Suppose inflation was 4% and has now fallen to 2%, it means that prices were increasing by 4% but that the rate of increase has now fallen to 2%. But remember, prices are still going up. We mention this elsewhere in the book because it is important that you understand what is happening to your money.

Pay rises

Because of inflation, trade unions have traditionally demanded an increase in pay to keep pace with inflation. They argue that if a worker earns £100 per week and the rate of inflation is 2%, then the worker needs a pay increase of 2% simply so to maintain their buying power. To receive a pay rise less than the rate of inflation is, in some people's eyes, actually a pay cut.

Calculating your pay rise

This is the same type of problem that we looked at earlier. Take the top line of your pay slip (the gross pay). Using the year-on-year inflation rate, find the new level of salary/wage that you need to achieve in order to stand still, economically speaking.

Let's say Mike works in a factory and earns £200 a week. If the year-on-year inflation rate is 4%, what does Mike need to earn next year in order to have the same buying power?

His new pay needs to be 104% of his old pay:

$$^{104}/_{100} \times £200 = £208$$

To have the same buying power, Mike needs to have his pay increased to £208 a week.

What is a ratio?

A ratio is a relationship between one quantity and another. We use ratios in many aspects of life, for example mixing concrete, making compost for the garden and in cooking. When you hear a builder, like UK televsion celebrity Tommy Walsh, talk of 'two parts sand to one part cement' then he is talking a ratio. It may be that those parts are shovelfuls or kilograms but it really doesn't matter – it is still a ratio.

Suppose Tommy wants to make a mixture that was 2 parts sand to 1 part cement. That means that whatever amount of sand he uses in the mix, he should always add half that amount of cement.

For example:

4 shovels of sand should be mixed with 2 shovels of cement;

8 kilograms of sand should be mixed with 4 kilograms of cement.

Ratios are written using a colon. For instance, the ratio 2 to 1 is written as 2:1.

Solving problems using ratios

Example 1

Imagine you have a compost mix for a lawn top dressing. It is made up of earthworm casts and sharp sand, in a 3 to 1 mix. There is a total of 10 000 kg of compost. What weight of earthworm casts was used in making the mixture?

The mix was 3:1, so we can think of there being 4 parts altogether and there is three times as much earthworm casts in the top dressing as there is sand. We can work out the weight of 1 part:

$$1000 \text{ kg} = 4 \text{ parts, so } 1 \text{ part} = {}^{1000\text{kg}}/_4 = 2500 \text{ kg.}$$

Since the ratio was 3:1, the amount of earthworm casts used was 3×2500 kg = 7500 kg. The remaining 2500 kg was made up of sharp sand.

Example 2

Concrete is mixed in the ratio of three parts fine sand to one part sharp sand to two parts cement. How much cement is used to mix 500 kg of concrete?

First of all rewrite the ratio in a more friendly form, 3:2:1. This stands for 3 parts fine sand to 2 parts cement to 1 part sharp sand. Altogether this makes 6 parts (3 + 2 + 1). So 6 parts is equivalent to 500 kg of concrete. So now we can work out what one part is worth:

6 parts = 500 kg, so 1 part = $^{500kg}/_6$ = $83^1/_3$ kg.

So two parts of cement must be $83^1/_3 \times 2 = 166^2/_3$ kg.

What is proportion?

Two quantities are said to be in proportion if they increase or decrease in the same ratio. For example, suppose potatoes cost 20 pence per pound. If I bought 2 pounds I would expect to pay 40 pence. I would also expect to pay 10 pence for $^1/_2$ lb (half a pound weight).

Income tax gives a good example of proportion. The more you earn, the more income tax you pay. The *proportion* of your taxable income that goes in tax is the same as for other people but the *amount* paid depends on your income.

It is not quite as simple as this example suggests, because there are bands of income which are taxed at different rates.

Journal	• Make notes summarising the main ideas covered in this chapter.

6　Do it Yourself

'Take care of the pence and the pounds will take care of themselves'
William Loundes

One-minute overview

Any book on mathematics for adults that does not look at 'do-it-yourself' is doing the reader a disservice. In our house, like millions of others, we save money by doing some jobs ourselves. DIY in the household is usually straightforward, but with some basic calculations you can save even more money.

In this chapter we will look at:

- the definition of perimeter;
- how to calculate the areas of shapes and the volumes of materials;
- how to find the circumference of circles.

Health warning
We strongly recommend that, unless you are qualified, you leave electrical jobs to the professionals. We also advise that you take the usual health and safety cautions.

From the beginning of January 2005 in the UK it became a legal requirement that all electrical work be carried out by a 'competent person'. This person must supply you with a certificate showing that a government-authorised electrical contractor –such as a NICEC contractor – has carried out the work. If you have one of these certificates *do not* lose it. You will need it when you come to sell your home. This law also applies to older properties that are being altered or having whole or part rewiring. The work will need to be inspected by the local authority. If you do not obey this law and subsequently an incident occurs where someone is hurt, you will be liable both under criminal and civil law. It really isn't worth the risk.

Perimeter and decorating

The perimeter of a shape is the distance all the way round its outside – by this we mean around the boundary of the shape.

Say you have a room in which you want to fix a new skirting board (a shaped board fixed at the base of the wall where it meets the floor). The perimeter of the room is the measurement to be made. The skirting board length is the distance around the boundary of the room.

To find the perimeter of a square, simply measure the length of one side and multiply by 4. The room here is rectangular. So the perimeter is calculated by measuring the width and the length and then using the formula:

$$\text{Perimeter} = (\text{width} \times 2) + (\text{length} \times 2)$$

You might not want the length of the door (and other things) included so that you can reduce the cost and waste of wood, so subtract this from the number the formula above generates.

However, a piece of good advice is to add a small amount, say 10%, to your answer. There is nearly always some wastage in cutting and this extra allows for this waste.

What about covering floors and walls?

To do this you need to know about area. One aspect of dealing with area is being able to communicate how big an area is. To do this we use standardised measurements. A square is a good standard to measure an area.

The square drawn here has each side 1 cm long.

It is a 1 centimetre square – [ ] it has an area of 1 cm^2.

As each side is 1 cm long it must be 10 mm long. So the area of a 1 centimetre square must be 10 mm by 10 mm, or 100 square millimetres (100 mm^2).

Most DIY firms now quote dimensions in millimetres rather than centimetres, so you will hear the term 'mill' being used for millimetres.

How to calculate areas

The rectangle drawn here has a length (the longest side) of 4 cm and a width of 3 cm.

Here is a bit of algebra for you. Don't panic, but there is a simple rule that your children will be learning at school to show how to calculate the area of a rectangle:

area = length × width   or $A = lw$

So the area of this rectangle is 4 × 3 or 12 square centimetres. If you look at the shape and count the small squares you will find that there are 12 of them. Each square has sides of length *1* cm.

Carpets

When dealing with carpet in rooms, you use the same approach. Carpets are usually be sold by the square metre (m^2). You will need to measure the length and width of the area you want to cover in metres to be able to calculate the area in square metres easily – remember, the mathematics will be exactly the same as above.

Walls and ceilings

Take the same approach as for calculating area for carpets by measuring the lengths and widths of any rectangles involved and multiplying the lengths by the widths. You will need to add the separate areas together to get the total area of the wall or ceiling. But don't forget to subtract the area of any cut-outs like doors and windows. Then you need to look at the coverage of materials such as paint. You will nearly always find a guide as to the area you can expect to cover, usually for the contents of the tin – for example, this paint will cover 25 m^2 when applied with a light brush.

Area in the garden

Use the same approach when working out areas in the garden. If you want to create a rectangular border or a rectangular lawn the measuring and calculation methods will be the same. Suppose you create a lawn that measures 10 m by 3 m. It will have an area of 30 m^2 so you will need to buy enough turf to cover that – plus a bit more to cover wastage.

If you are going to apply a dressing you need to look into the ratio of application per unit area which will be printed on any packets you buy – unless you decide to make your own! Suppose you decide that you need 500 grams per square metre and that a dressing mixture made up of 3 parts earthworm casts to 1 part sharp sand will do the job. How much top dressing do you need to make?

First of all you will need to decide on the weight of 1 part of top dressing – for the purposes of this calculation let's assume that 1 part of top dressing weighs 250 g. You need 500 g per square metre so that's 2 parts of dressing for every square metre. The area of your new lawn is 30 square metres so you will need $2 \times 30 = 60$ parts of top dressing.

Since each part is 250 g you will need 60×250 g = 15 000 g, or 15 kg, of top dressing. You can then work out how much sand and earthworm casts to mix together.

You may think that this is trivial given the concerns mentioned earlier in the book but, with respect, that does somewhat miss the point. One of the recurring themes throughout this book is the need to constrain expenditure but still enjoy a certain quality of life. We know of people who have over-ordered on quantities and have been left with materials they can do nothing with – what a waste of money!

Some years ago we heard of one mathematics teacher and her doctor husband who said that they did not bother working out areas of walls to find out how many tiles they needed for their bathroom – we wonder if they could?

Could you? To work out how many tiles needed for a bathroom, you need to work out the area of the walls, and most walls are rectangles. You would then choose your tiles and work out the area of each tile from the measurements on the box in the DIY store. You would then divide the wall area by the area of one tile and that would show the minimum number of tiles needed – of course, there will be cutting and wastage but you just add a sensible extra factor. Maybe that's the bit that the couple mentioned above didn't fancy. Anyway they just estimated how many tiles they would need and ran out about four times. So they had to return to the DIY store each time to buy another box hoping that they would have enough this time – a crazy waste of time, when a simple calculation would have been close the first time.

You may agree with them on their chosen 'strategy' but while the tiles you first bought may still be in stock, the tint or colour may be slightly different (they are made in batches) or you may be really unlucky and find that they're out of stock and the range has been discontinued. These are risks you take if you don't know how, or can't be bothered, to calculate the area and determine how many tiles you need to buy in one lot. How many unfinished bathrooms exist simply because people could not calculate the area of their walls?

Journal	• Explain the concept of area in your own words.

Making money from DIY

If you are interested in making money to pay off debt then this is certainly an area worth considering. Here are some suggestions.

• If you have a spare room, decorate it and rent it out. A spare room could be worth £50 per week from a lodger, depending on where you live. £50 per week means you will earn £2600 per year. Remember that Jeffrey Archer, the author, used what he had (a pencil) to create wealth (by writing books). Here you are using the same principle – use what you have got to create wealth for you and your family. In terms of decorating the room, the experts we asked recommended that the walls are painted a neutral colour, for that read magnolia, and that the carpets are a terracotta type colour. This gives the room a sumptuous look and means that the carpet contrasts well with the walls – it also makes the room feel warm, and that the carpet does not show many marks.

• Gardeners can earn money by providing simple gardening services during weekends for retired people. Look for elderly ladies in your area and provide a service cutting lawns. All you need is a mower and the means to transport it. We know of one successful small business in Somerset where the owner, himself retired, provides this service. He charges £5 for mowing a lawn and can do about 8 lawns a day. This is £40 per day just for the lawns. Add in cutting back trees, tidying hedges and weeding and you could easily increase this to £15 per visit. This gives you a possible £120 per day. Let's say you did this every Saturday – after ten Saturdays you will have £1200, and all you have done is use the family lawn mower. Again, think *safety* – for a small cost you can buy a circuit breaker at your local DIY store. If you are using an electric lawn mower and cut the cord by accidentally running over it, this breaker will cut the circuit and possibly save your life. When we looked around we found that they cost about £15 ($30 or €25), a sum that is certainly not worth risking your life for.

- Security is another issue for many elderly people. Here is another idea for using your DIY skills for profit. Price the cost of door security chains in your local DIY. Then add 25% to this to cover the cost of travel to buy the chains. Then double this cost. This is your minimum selling price. Have a leaflet made and distribute it around your area where older people live. Offer to supply and fit security chains and you should get many takers. Let's look at the figures. Say a chain costs £5, you aim to buy 10 – your initial outlay is £50. Add 25% (£12·50) to cover time and travel. Don't forget that you'll need to print the leaflets and that will cost something so add an amount to cover this – you should be able to run it up on a word processor and print, say, 100 for about £10. The total cost is now £72·50 so you double this to £145. Now subtract the initial £72·50 cost to leave £72·50 – that is your potential profit on just 10 houses i.e. £14·50 per house. If you did 100 houses you would make £725 profit. You should be able to fit a chain in less than 10 minutes so you should easily be able to do 20 houses on a Saturday. That is £105 *profit* for a few hours work on a Saturday – and you have not had to give up the day job.

- People who are a little more adventurous could offer
decorating services – but to do this you *must* be good. If
you do decorate someone's house we suggest you use this as
your calling card, or better still decorate your own home and
use that as a calling card. By 'calling card' we mean use it as a
mechanism to generate other potential jobs. One way of
doing this is to photograph the rooms before and after. A
handy tip is to make the 'before' photographs dark and
ensure that the rooms are untidy. For the 'after' photographs
set the room in an attractive way so that it is light and airy.
Remove all personal belongings and show the room in the
same way that a room in a builder's show home would be set.
If you are not sure how to do this, spend a Sunday afternoon
travelling round new estates and see how it is done.

Whatever your chosen DIY activity, you *must* make sure
that you do a great job. If your work is good, then your
customers will brag about how they have used your services
and how secure you have made them feel. Remember that
people buy feelings and emotions – here they are buying
security. If you do a bad job they will also tell their friends
how poor you are at your job. On average, according to
research we have seen, a happy customer will tell about nine
other people. An unhappy customer will tell about 27 other
people. So make sure you leave them happy.

Can you see the potential here? It is a strategy that you can
use and develop to make money and help you and your
family to create wealth and eliminate debt from your lives.

Don't be shy about asking for referrals. Ask your customers
to tell you who else might benefit from your service. Then
approach that potential customer by saying 'Mrs X asked
me to do a job for her and was so pleased with the work
that she suggested I contact you because you might want
the same job doing.'

A word of caution about property development

In the examples above we have made suggestions to help you make an extra income. Some people move on and go into property development. Sarah Beeny on Channel 4 TV in the UK presents an excellent programme called *Property Ladder*.

Ms Beeny is a professional property developer – that is her day job – yet many first timers in the programme choose to ignore what she is saying. If you do want to move into property development then please do your homework. This is called desk research. It takes only your time, not your money and it is vitally important – it can save you losing everything.

Sarah Beeny has a number of excellent books on the market. They really are worth buying and reading. We have been appalled at tales of how people have bought property 'unseen'. This is crazy and a sure way to undo all the good work you have done in getting yourself out of debt. The purpose of this book is to look at the mathematics that many adults need – we are not going to discuss the property market but we will say this, *be careful.* From 2005 onwards, the property market in the UK is unlikely to grow at the rate it has since the year 2000. Experts are split about how it will develop and frankly you need to take a view. In the *Mr Educator* office we believe the market will probably go sideways. By this we mean that the market will grow but in single digit figures for a few years. We do not agree with the gloom merchants who say there is going to be a crash. Towards the end of 2004 the headlines stated that there is a 30% chance of a crash – that means there is a 70% chance that there will not be a crash. The trouble is that the 70% figure doesn't make such a good headline.

But we say, again, be careful because you can lose money on homes as well as make it. We did see one pair of sisters

on Sarah Beeny's programme who lost control of the spend
on their property and were heading for a £34 000 loss – it
can and does happen. Remember the words assets and
liabilities – any property development should be an asset
not a liability. The American author Robert Kiyosaki writes
'assets feed you, liabilities eat you'. (The *Rich Dad, Poor Dad*
series is well worth your time in reading. Although it is
American-based, and clearly UK tax laws are different, there
is a huge amount of excellent advice in these books.)

Journal	• What services could you offer? Do a SWOT analysis. Identify your Strengths and Weaknesses, what Opportunities are present in your area and what are the Threats to your business idea?

Cakes and circles

Another good small-time enterprise is cake making. We
recently met one lady, Anwen, who does excellent business
in Wales making and selling wedding cakes. There is some
simple mathematics at work here. Circular cakes will need
ribbons around them and there is a simple way of finding
the length of ribbon needed.

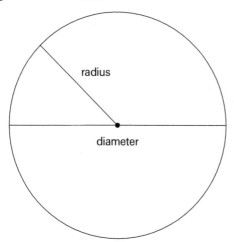

First of all it is important to understand the names of the parts of the circle.

Circumference: this is the outer perimeter of the circle and is a continuous length.

Diameter: this is a straight line through the centre of the circle from any point on the circumference to another point on the circumference.

Radius: this is a straight line from the centre point to any point on the circumference. It is always half the length of the diameter.

It is also common to use these words to represent the lengths involved – so we speak of a circle having a radius of 2 cm.

Circles and π

When you were at school you may have done an experiment in which you had to place string around circular objects. This gave you a rough idea of the measure of the circumference of a circular object. You will have then used some more string to measure the diameter and then find how many times the diameter divided into the circumference. The idea was to introduce you to pi –an important number in mathematics, represented by the Greek letter π. Most calculators have a π button – very useful for doing circle sums.

If you take any circle and divide its circumference by its diameter, you always get the same answer, namely 3·141592654 (to 9 decimal places). Try it for yourself – the value of pi has never been fully determined, although it has been worked out to millions of places. There are two Russian brothers (the Chudnovsky brothers) who live in the United States who do nothing else but work out places for pi – quite why no-one knows but that is the beauty of living in a free country.

We can use pi to work out the circumference of a circle. There are two formulae for finding the circumference, but really they are the same because a radius is always half of the corresponding diameter:

$$C = \pi d \text{ and } C = 2\pi r$$

where C is the circumference, d is the diameter and r is the radius. Remember that when two letters are placed next to each other, as in these formulae, it means multiply the variables.

Generally, we use the first formula when we know the diameter of the circle. Also, if you don't use the π button on your calculator, we use the value 3·142 for π.

Example 1: Find the circumference of a circle of diameter 5 cm.

$$
\begin{aligned}
C &= \pi d \\
&= 5 \times 3142 \\
&= 15{\cdot}71 \text{ cm (2 d.p.)}
\end{aligned}
$$

Example 2: Find the circumference of a circle of radius 4 cm.

$$
\begin{aligned}
C &= 2\pi r \\
&= 2 \times \pi \times 4 \\
&= 25{\cdot}13 \text{ cm (2 d.p.)}
\end{aligned}
$$

So in estimating the length of ribbon around a cake, you need to use three times the diameter (or six times the radius) and then add a small amount. This will mean you can buy the right amount of ribbon.

Just in time (JIT)

You need to plan your work so that you do not need to hold vast quantities of stock. Stock costs money and this is money that you cannot use when it is locked up in the form of stock. In the example of the wedding cake manufacturer above, Anwen would order enough ribbon for the cakes she was making and then go back and order more when needed, rather than order a huge amount to keep and use when needed. This frees up your cash flow (which we will discuss at the end of the book) and means you are more flexible. It is a technique used in manufacturing companies and is commonly known as *just in time*.

In other words materials are ordered and delivered just when they are needed. This needs good planning but it will save you money.

Journal	• Explain the work on circles in your own terms.

The area of a circle

There is another formula for finding the area of a circle – it is

$$A = \pi r^2 \text{ (remember, } r^2 \text{ means } r \times r)$$

Example 1: Find the area of a circle of radius 5cm.

$$
\begin{aligned}
A &= \pi r^2 \\
&= \pi \times r \times r \\
&= 3{\cdot}142 \times 5 \times 5 \\
&= 78{\cdot}55 \text{ cm}^2
\end{aligned}
$$

Remember that area is always in square units – here it is square centimetres.

Example 2: Find the area of a circle of diameter 30 cm.

If the diameter is 30 cm then the radius must be 15 cm.

$$
\begin{aligned}
A &= \times r^2 \\
&= \pi \times r \times r
\end{aligned}
$$

$$= 3 \cdot 142 \times 15 \times 15$$
$$= 706 \cdot 95 \text{ cm}^2$$

This is the technique to use whenever you need to know the area of a circle – as when you have to add top dressing to a circular lawn.

Volume and conservatories

You may be amazed to find that in many areas of the UK planning permission for a conservatory depends on the volume of the proposed construction. (Conservatories are very popular in the UK. They are largely glass-built rooms that are extensions to the main property and allow people to enjoy being closer to the garden without the infamous inclement weather affecting them.)

For example, in one area in the south west of England you can build a conservatory without planning permission so long as the volume is less than three cubic metres.

What is a cubic metre?
Think of a box with each side a metre long – then you have your cubic metre.

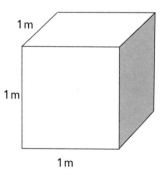

To find the volume of a cuboid (a box) multiply the length by the width by the height. So when it comes to determining the volume of your conservatory, you need to make a good estimate of the length, height and width and then multiply them together.

Health warning
We have seen some conservatories that simply do not sit well on the back of the house to which it is attached. The conservatory has to be in proportion to the rest of the house. If it is too large, it will simply look wrong. In the famous words of the Prince of Wales, it will look like a 'monstrous carbuncle on the nose of a well-loved friend'.

We suggest that the height and width of the proposed conservatory should be less than one-third of the overall dimensions of the side of the house to which it is to be attached.

However, it is your house so you decide!

Now, brace yourself – we are going to talk algebra.

7

Algebra, Business and Spreadsheets

'Winning, that is why we are here'
Sign outside the changing rooms at Twickenham

One-minute overview

Many adults hold up their hands in horror when the word 'algebra' is mentioned, but this does not have to be the case. With a bit of careful thinking you will be able to deal with the algebra we will discuss here. But why are we discussing algebra here? Didn't all that end when we left school? Well not quite. You see, every time you create a spreadsheet or use a formula you are using algebra, and that is the purpose of this chapter.

In this chapter we will look at:

■ using spreadsheets;
■ brackets and algebra;
■ how to substitute into formulae and make sure you get the right answer;
■ solving simple equations.

What is a spreadsheet?

	A	B	C	D	E
1		Cell B1			
2			C2		
3	Cell A3				
4				...D4	

In this chapter we can give only a brief introduction to spreadsheets but it should be enough to get you going. We are going to assume that this is the first time you have used a spreadsheet. It is a brilliant tool and can save hours.

Each *column* on a spreadsheet is labelled by a letter and each *row* by a number. The rectangles in the body of the spreadsheets are called *cells*.

Each cell is given its own 'address' defining exactly where a particular cell lies in the spreadsheet. The cell where column A and row 3 intersect is called A3. A continuous rectangular group of cells is called a *range*. A range is defined by the cells at its upper left and lower right corners. In the example above the range represented by the shaded area is called C2:D4.

This is what a typical spreadsheet looks like on screen. There are many spreadsheet programmes available but they look and work pretty much the same – and it helps if you understand a little about algebra to get the best out of them.

Screenshot courtesy of Microsoft Corporation

Whichever cell is highlighted (with a dark border) is called the *active* cell. You can move around a spreadsheet using the arrow keys on the keyboard or the standard keyboard

shortcuts, or clicking around
with the mouse.

If you need to move a long way,
the Go To ... feature might be
useful – click Edit and you'll find
it in the drop-down menu. Just type
in the address of the cell you want
to move to and click OK.

Screenshot courtesy of Microsoft Corporation

If the columns are not wide enough, or are too wide, or the
rows not deep enough you can adjust their size by clicking

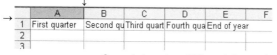

	A	B	C	D	E	F
1	First quarter	Second qu	Third quart	Fourth qua	End of year	
2						
3						

Screenshot courtesy of Microsoft Corporation

and holding on the line between the two letters or two
numbers and dragging to fit the text. Column A has been
widened in this screenshot.

Types of data in spreadsheets

Any spreadsheet cell can contain a label, a numeric value or
a formula:

- a label is made up of text or numbers, which are *not* being
  used in calculations. A common use for labels is as column
  headers or row names;

- a value is a piece of data, like a number or a data, which can be
  used in calculations;

- a formula is an algebraic expression which defines a
  calculation to be done on one or more cells.

First of all we will do some simple examples before moving
onto the algebra.

Example 1: Adding

Here we have a set of numbers (values) and want to add them and put the total in cell A11. You could just do this mentally but it can be tedious, especially if there is a lot to do.

You could add them up by simply typing the formula =A1 + A2 + A3 ... but that's pretty tedious too.

A quicker way, making more use of the spreadsheet's power, is to use a formula to work on the range of numbers. You click in cell A11, where the answer is going to be, and enter the formula =SUM(A1:A10).

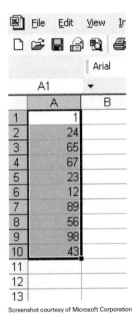

An even quicker way is to use the SUM button on the toolbar – it looks like Σ. This is another Greek letter called 'sigma' and here it means 'add up'. You highlight the column by clicking in A1 and dragging down – this defines the range to be worked on. Now single click in cell A11 and then click Σ.

Screenshot courtesy of Microsoft Corporation

However you do this you should find that the answer is 478.

Example 2: Averaging

The average for the data in a range of cells can be found in different ways too.

If you want to work out the average of the fourteen numbers in this spreadsheet and put the answer (75) in cell B15, you could make B15 the active cell and enter this formula =SUM(A1:A14)/14.

It's even easier to enter the formula =AVERAGE(A1:A14), you don't even have to count the number of items to divide by.

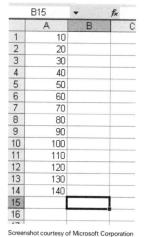

	A	B	C
1	10		
2	20		
3	30		
4	40		
5	50		
6	60		
7	70		
8	80		
9	90		
10	100		
11	110		
12	120		
13	130		
14	140		
15			
16			

Screenshot courtesy of Microsoft Corporation

How to use algebra to make formulae

In this example we are going to calculate the gross salary, the tax payable (at a rate of 23%) and the net salary for a group of six employees. We know that:

- gross salary = hours worked × hourly rate
- tax payable= gross salary × 0·23
- net salary = gross salary − tax payable

The spreadsheet below has all the primary data.

	A	B	C	D	E	F
1	Name	Hours worked	Hourly rate	Gross salary	Tax payable	Net salary
2	Tony Sulliton	35	40			
3	Emma Buying	22	38			
4	Peter Murrow	1	25000			
5	Alison Murrow	0.5	50000			
6	Elaine Field	33	38			
7	Neil Whilhams	55	12			
8						
9						

Screenshot courtesy of Microsoft Corporation

Tony is a top executive and earns £40 per hour. Elaine and Emma run departments and are the bedrock of the company; Neil is the junior and has to work a lot of hours to make money. The spreadsheet needs some formulae now.

The gross salary for Tony is calculated using the formula =B2*C2. Note that * is the symbol for multiplication in spreadsheets. Emma's gross is calculated using the formula =B3*C3, and so on. The spreadsheet below shows the formulae used.

	A	B	C	D	E	F
1	Name	Hours worked	Hourly rate	Gross salary	Tax payable	Net salary
2	Tony Sulliton	35	40	=B2*C2		
3	Emma Buying	22	38	=B3*C3		
4	Peter Murrow	1	25000	=B4*C4		
5	Alison Murrow	0.5	50000	=B5*C5		
6	Elaine Field	33	38	=B6*C6		
7	Neil Whilhams	55	12	=B7*C7		
8						

Screenshot courtesy of Microsoft Corporation

Here is the spreadsheet when all the gross payments have been calculated.

	A	B	C	D	E	F
1	Name	Hours worked	Hourly rate	Gross salary	Tax payable	Net salary
2	Tony Sulliton	35	40	1400		
3	Emma Buying	22	38	836		
4	Peter Murrow	1	25000	25000		
5	Alison Murrow	0.5	50000	25000		
6	Elaine Field	33	38	1254		
7	Neil Whilhams	55	12	660		
8						

Screenshot courtesy of Microsoft Corporation

We will make a simple assumption about tax payments. We have calculated income tax on the gross pay at 23%. In reality there are tax allowances and several tax bands across the salary range, which the spreadsheet would be able to handle easily – here we have kept it simple.

Tony's tax bill is 23% of his gross salary, that's 0.23 of the value in cell D2. So the formula to use for him is =(D2*0.23).

The net salary will be the gross salary minus the tax payable – for Tony this will be =D2−E2.

The spreadsheet below shows the formulae used for tax payment and net salary calculations.

	A	B	C	D	E	F
1	Name	Hours worked	Hourly rate	Gross salary	Tax payable	Net salary
2	Tony Sulliton	35	40	=B2*C2	=D2*0.23	=D2-E2
3	Emma Buying	22	38	=B3*C3	=D3*0.23	=D3-E3
4	Peter Murrow	1	25000	=B4*C4	=D4*0.23	=D4-E4
5	Alison Murrow	0.5	50000	=B5*C5	=D5*0.23	=D5-E5
6	Elaine Field	33	38	=B6*C6	=D6*0.23	=D6-E6
7	Neil Whilhams	55	12	=B7*C7	=D7*0.23	=D7-E7
8						

Screenshot courtesy of Microsoft Corporation

And here is the spreadsheet showing the final values.

	A	B	C	D	E	F	G
1	**Name**	**Hours worked**	**Hourly rate**	**Gross salary**	**Tax payable**	**Net salary**	
2	Tony Sulliton	35	40	1400	322.00	1078.00	
3	Emma Buying	22	38	836	192.28	643.72	
4	Peter Murrow	1	25000	25000	5750.00	19250.00	
5	Alison Murrow	0.5	50000	25000	5750.00	19250.00	
6	Elaine Field	33	38	1254	288.42	965.58	
7	Neil Whilhams	55	12	660	151.80	508.20	
8							
9							

Screenshot courtesy of Microsoft Corporation

Brackets and algebra

Brackets are important in algebra. Basically it means that you need to work out what is inside the brackets first. Suppose you wanted to know only the net salaries to be paid in the examples used above – there's no need for all these columns. If you think about it all you need is the hours worked, the hourly rate and the tax rate. You could then use a formula like this =(B2*C2) - (B2*C2*0.23) for Tony, but that works only if you know how brackets work.

And if you really think about it you could get away with =B2*C2*0.77, but only if you clicked with the percentages we did earlier on.

Equations

We have left the best to the last in this chapter. Equations cause fear in many people so we have included some simple algebraic equations here so that:

● you can use them in your work, say on spreadsheets;

● you can help your children when they are stuck on algebra homework.

In truth, the latter is where most people will come into contact with equations. But equations are used in the workplace and you should not be surprised to come across them. You might even find ways of putting them to use in everyday life.

As far as children are concerned, one of the benefits of learning to solve equations is that it helps train them to think ahead. What are you doing when solving an equation? You are using a strategy to achieve a result. This is exactly the type of higher-level thinking that is in desperately short supply in industry. There are new types of jobs in industry that can be described as involving *analytical thinking* or *critical thinking*. The popular name for these is 'troubleshooting' where people create theoretical models to solve problems and then apply them in the real world.

Algebra helps your child, and you, to develop this higher-level thinking and that is why it is important in general – but mathematics is also a subject worthy of study in its own right and we must never lose sight of that.

An image to use when working with equations

First of all, think about an equation as an old-fashioned balance. By this we mean the type of scale that grocers used to weigh vegetables – the kind of scale where vegetables were placed on one side and weights placed on the other side to balance.

Treat an equation in the same way – whenever you do something to one side of an equation, you must do *exactly* the same to the other side. Otherwise it becomes a grocer's scale that is unbalanced.

Let's look at some simple examples.

Example 1
Think about the equation $x + 7 = 10$. You should read this as an 'unknown number plus 7 gives an answer of 10'. It's a bit more obvious then that the value of x must be 3, simply because $3 + 7 = 10$.

However there are two issues here. The first is that your children will not have the adult sophistication to 'see' this immediately. The second is that the most important learning outcome here is that you and your children should *understand* the process used – this is the process used to solve many equations.

Comment	• Just a piece of advice – don't expect your children to 'see' the answer here. Many children simply will not understand what is happening. This is normal development, but when parents show exasperation and bewilderment it simply leads to children feeling insecure and this can cause neurotic feelings towards mathematics. Tr y to keep your feelings private – be patient and positive towards your children's progress.

The mathematical way to solve equations such as $x + 7 = 10$ is firstly to recognise that it is a value for x that we are after. You then 'undo' the equation by taking 7 away from both sides:

$$x + 7 - 7 = 10 - 7$$

Look at what we have done here. All we have done is put -7 on *both* sides of the equation. Look at the left-hand side – this now reads:

$$\text{unknown number} + 7 - 7$$

The 'unknown number $+ 7 - 7$' must just be the unknown number. If I have an orange and two apples and I take two apples away, I must be left with the orange. The same approach works in algebra. By taking away 7 we are left with the x.

On the right-hand side when we work out '$10 - 7$' we are left with 3. So the final answer is $x = 3$.

This series of steps can be written in this format:

$$x + 7 = 10$$
$$x + 7 - 7 = 10 - 7$$
$$x = 3$$

Notice how all the equal signs are lined up – this is something to be encouraged because it makes the mathematics easier to read and follow.

Example 2
Solve the equation $x + 19 = 7$

Undo the +19 by taking 19 from both sides

$$x + 19 - 19 = 7 - 19$$
$$x = -12$$

What about equations with an *x* on both sides?
First of all don't panic. There really is no need to reach for the scotch. Let's look at an example.

$$x + 7 = 2x + 2$$

The first strategy here is to get all the algebraic variables (the letters) on one side of the equal sign and the numbers on the other side. In other words, we are *simplifying* the equation by writing it in a form we recognise.

Take the smallest variable (x is smaller than $2x$ so that's x in this case) away from both sides:

$$x + 7 - x = 2x + 2 - x$$

Now tidy this up:

$$7 = x + 2$$

Now take the smallest number away from both sides.

$$7 - 2 = x + 2 - 2$$

And tidy this up

$$5 = x$$

So $x = 5$. Simple!

There will be many examples in your children's maths books to work through together and develop their knowledge – all we are trying to do here is get you out of trouble.

A further source of information, hints and practice is given on the BBC Learning website at www.bbc.co.uk/learning

Part Two
Handling Data

Communicating Data

> *'Be happy while you are living because you are a long time dead'*
> Scottish saying

One-minute overview

Communicating data means transferring information through the use of numbers. This is something that is generally done incredibly badly and there is little sight of it improving. Some ways of communicating data are very clear, but there are also ways that are deliberately meant to mislead.

There must be a set of rules that we can use to guide what we are doing – and there is. The mathematician David Targett has devised seven rules of communication and they are excellent measures to compare your work against.

In this chapter we will look at:

■ how to present data in tables;
■ how to make the information we are trying to communicate very clear.

Presenting information clearly

It's surprising that there should have to be a *Campaign for Clear English*, but there is. It is an attempt to remove jargon from the language to make it accessible to all who wish to understand what is meant to be communicated. Yet there is no equivalent in numbers. Communicating data can generate the most obscure and impenetrable morass of information, which then becomes ignored. If we don't make data accessible, it not surprising that people 'vote with their feet' and ignore the work of mathematicians.

You may generate data in the course of your own work, or you may need to use data in business documents, management information systems or accounting reports that are produced in the course of your work. The purpose of communicating data is to communicate overall *patterns* that can help in making decisions. The overview in data handling can be represented by *QCAI*.

- **Q** is for query: Data handling starts with asking a question. You need to ensure that you are asking a question for a purpose and not simply engaging in data collection for the sake of it. Nobody collects data for the sheer joy of collecting – well that is not quite true. The scientist John Dalton is said to have collected rainfall data every day for the best part of 60 years in Manchester, and he did nothing with the data. But then he was an inordinately boring man. Normally there is a question to be answered. For instance, are house prices higher on average now than compared with five years ago? This gives a purpose to collecting data.

- **C** is for collect: This is the stage in which we gather information, in the form of questionnaires or data collection sheets. This is incredibly easy to get wrong and must be organised with extreme care. One local council in the south of England constructed a questionnaire that was to be used to plan for a community recreation facility. They only received a 10% return from the questionnaires distributed. Of those who responded, 95% were in favour and this 'majority' was then used as justification for spending tens of thousands of pounds of public money. The same council was staggered when the new facility was hardly used and they had to bail it out financially.

- **A** is for analyse: This is the stage in which the graphs and charts are drawn. These give an overview of the data.

- **I** is for interpret: This is the stage in which we look at what the figures are telling us.

A word of warning, data handling goes wrong when *any* of these broad stages are short-circuited.

But even before we get to these stages, we need to go through Targett's seven rules for data presentation.

Rule 1: round numbers to two effective figures

Given a problem such as $29.8 \div 51$, most people would either give up or approximate and work out something like $30 \div 5$ and get the answer 6. In other words, this is something that most of us intuitively do anyway – all Targett does is formalise the process. There are, however, circumstances when this level of approximation is not appropriate, for instance when dealing with quantities of medicine to be administered to a patient.

The general rule is that if the last figure in a number is 5 or more then we round up to the next interval. I know we have covered this before in Chapter 4 (the 'five rule') but a little more won't hurt.

For instance, if we had to round 36 to two effective figures, it means that we need to round it to the nearest 10. Think about how it looks on a number line

30_____36_____40

Clearly 36 is closer to 40 so we would round 36 to 40 (to two effective figures).

What about 34?

30_____34_____40

34 is closer to 30 than to 40 so 34 would round to 30.

What about 35?

This is where the 'rule' kicks in.

30_____35_____40

35 is the same distance away from 30 as from 40. So we use the rule that if the last number is 5 or more we round up to the next interval. So here we round 35 up to 40.

The equals sign is common in mathematics expressions but sometimes its use is not appropriate. $\approx$ is the symbol we use for approximately equal. So, after rounding to two effective figures, we say 35 $\approx$ 40 and this reads as '35 is approximately equal to 40'.

Some students will write 35 = 40 after this rounding. This is not true – they are not *equal*. It is a good example of poor communication. What the student is trying to say is 35 is approximately equal to 40 but they have used the wrong symbol and so communicated wrongly.

What about rounding to two decimal places?

Here the two *effective* figures are the two figures just after the decimal point. So we can ignore the numbers in front of the point and just consider the decimal part of the number. Take 2·459 for instance. Clearly the number we need is either 2·45 or 2·46 and we have to decide just which.

2·45_____2·459_____2·46

A number line makes this clear. You can work without the inconvenience of drawing number lines. Just concentrate on the required number of (effective) figures after the decimal point and apply Targett's rule 1. Therefore, to round to two effective figures, look at the *third* number – if that number is 5 or more, then round upwards, otherwise round down.

Rule 2: Reorder the numbers

It is far easier to understand what is happening with a set of numbers when they are written in order of size. Just to show what we mean, look at this set of data:

23·4, 34·4, 34·5, 23·5, 21, 19, 12, 16, 18, 24, 25·4

Confusing isn't it. Try reorganising this list in order of size and look at the difference:

12, 16, 18, 19, 21, 23·4, 23·5, 24, 25·4, 34·4, 34·5

We have separated the items of data with commas. This is fine in the UK but can cause problems elsewhere. In continental Europe, for example, the comma is commonly used to represent the decimal point – so 1,2 reads as 'one point two', another an example of poor communication like the resulting list:

12, 16, 18, 19, 21, 23,4, 23,5, 24, 25,4, 34,4, 34,5

Rule 3: Swap rows for columns

Look at this calculation: $123 - 34$

Now compare this with:
$$\begin{array}{r} 123 \\ -34 \\ \hline \end{array}$$

It is far easier to see what to do and how to work out the answer. Children in school are encouraged to set out subtractions in this way.

So why, when we are entering data in a table, do we present the data horizontally rather than vertically? It is far easier to see any patterns and trends when the data is presented vertically.

Rule 4: Use summary measures

Summary measures provide a way of describing averages. For example, it is easy to see how far each item of data deviates from the average value (mean) when presented in a certain way.

For instance, take the numbers 1, 2, 3 and 4. Now work out their average. This is the total divided by the number of items, or $10 \div 4 = 2.5$. Now subtract the mean from each number to work out the difference to get -15, -0.5, 0.5, 1.5.

Now let's look at this data again, but presented in a table.

Item	Difference from mean
1	-1.5
2	-0.5
3	0.5
4	1.5

By making a simple table like this, we can clearly see how far away each item of data is from the mean. But there is one problem – when you add up the deviations they cancel out. Try it and you will see what we mean. We need to be able to add the deviations so that they do not cancel each other out. This will give us a real estimate of the variability of the data.

There are two ways forward – *absolute deviation* and *variance*.

Absolute deviation

In this method we ignore the negative signs and treat all values as positive. If you put two vertical lines either side of the calculation, like this $|3 - 0.5|$, you get a measure called the absolute deviation. If the absolute deviation is divided by the number of scores then the result is the mean absolute deviation.

Variance

An alternative is to square each of the deviations – this has the effect of eliminating the negatives. Think about it – what happens when -1 is squared? You get 1.

So by squaring and eliminating the negatives we are ridding ourselves of the problem of the deviations being zero. The process now becomes:

- calculate each deviation from the mean;
- square each deviation;
- add up the squared deviations;
- divide this total by the number of items of data to get the mean of the squared deviations.

This mean of the squared deviations is called the *variance*. This figure is useful because it indicates the average variability of the scores around the mean, when they are expressed as squared deviations. The variance is also a good measure of spread because it is larger when the data are spread out far and small when the data are closer together.

Standard deviation

The variance is a measure of squared deviations from the mean, so it cannot tell us the distance from the mean in a frequency distribution. In order to find this deviation from the mean we need to find the square root of the variance – this is called the *standard deviation*.

To find the standard deviation therefore we need to:

- calculate each deviation from the mean;
- square each deviation;
- add up the squared deviations;
- find the square root of the total.

The standard deviation indicates the spread of the data about the mean. Generally, most of the results will lie within $+/-1$ standard deviation from the mean.

There is a potential problem in dividing the number of items of data. The number of items of data is represented by N.

You should only divide by *N* when:

- the values in the distribution represent the whole of the population;

- the values in the distribution represent a sample from the population and we are only investigating the variation and the standard deviation within the population itself.

When investigating the variation and standard deviation within the population, by taking a sample of size *n*, you should always divide by $n-1$. The reasons for this are complex and beyond the scope of this book. When you find a standard deviation from a sample this is called the *sample standard deviation*.

Rule 5: Use clean tables

This is good advice for any aspiring mathematician. You should leave as little white space and as few grid lines as possible in any table. This is contrary to what many teachers tell students. Many people have the view that lots of white space makes things easier to read, but putting lots of gaps in a table makes it hard to see patterns and it is this pattern spotting ability that is vital. The same is true for grid lines, whether horizontal or vertical. They should only be used to separate data that is different in context (look at the simple table in Rule 4).

Rule 6: Make labelling clear but unobtrusive

If you are labelling data make sure the system you use is clear. The last thing to do is confuse the reader by presenting graphs and charts that are so messy that it is difficult to understand the message they are meant to convey.

Rule 7: Make a written summary

This can be really helpful in communicating the meaning of the data. Your statement should draw immediate attention to the main features of the data. You need to keep it short and sweet, and it should deal with the main patterns shown in the data. It should not consider points of detail or oddities, and it most certainly should not mislead.

Lawler's addition to Targett's rules

To paraphrase George Orwell, you should break any or all of these rules rather than produce a set of data that is 'barbarous'.

Journal	• Construct a diagrammatic display about Targett's rules and Lawler's addition.

Politicians and lies and statistics

You must have heard the oft-quoted phrase 'lies, damn lies and statistics'. This is something that politicians are familiar with only too well. Certainly in the UK, the accepted norm of behaviour is that politicians should not lie to the electorate and, by and large, this is what happens. However, a politician can distort data patterns and still not be lying. Bold claims can be made while still, strictly speaking, be telling the truth. When the general public does not understand the manipulation that is going on, they can be misled. More often than not, one feels that voters can recognise that something untoward is occurring. This increases the feeling of isolation that the public generally feels with the world of politics and frankly is, in our view, one of the reasons why so many people are disengaging with politics. A good example is where a major UK politician is reported to have said that pensions need to go up. How do we interpret this? He did not say that if his party was in power then it would put pensions up, but some people may have interpreted it in that way. In that sense he didn't lie, but it demonstrates the care that should be taken in interpreting what politicians, and others, say. As one lady told us, 'We know we are being misled, we are not stupid'. How true!

Look at this table as an example. This table shows the per capita change in growth of purchasing power standards in the EU between 1996 and 2001. It is measured in percentage points and is for the 15 European regions.

Region	Change per capita GDP
Iperios (EL)	13
Trento (IT)	−10
Luxembourg	33
Valle d'Aosta (IT)	−16
Inner London (UK)	32
Leipzig (Ger)	−11
Southern and Eastern (IE)	27
Berlin (Ger)	−15
Cumbria (UK)	−13
Beds and Herts (UK)	13
Detmold (Ger)	−10
Peoloponisos (EL)	14
Madeira (PT)	16
Schlieswig-Holstein (Ger)	−10
Aland (FI)	16
Border, Midland and Western (IE)	16
Hamburg (Ger)	−11
Hannover (Ger)	−13
Koln (Ger)	−14
Berks, Buck and Oxford (UK)	22

These are actual figures by courtesy of the EU and show a five-year performance of what are called NUTS2 regions. But they are not the easiest of data to interpret. (To be fair to the EU officials, we have altered the way the table was presented to make a point.)

Let's apply Targett's rules here. In fact, most of the rules have been applied.

Region	Change per capita GDP
Luxembourg	33
Inner London (UK)	32
Southern and Eastern (IE)	26
Berks, Buck and Oxford (UK)	22
Madeira (PT)	16
Border, Midland and Western (IE)	16
Aland (FI)	16
Peoloponisos (EL)	14
Beds and Herts (UK)	13
Ipeirios (EL)	13
Trento (IT)	−10
Detmold (Ger)	−10
Schlieswig-Holstein (Ger)	−10
Leipzig (Ger)	−11
Hamburg (Ger)	−11
Hannover (Ger)	−13
Cumbria (UK)	−13
Koln (Ger)	−14
Berlin (Ger)	−15
Valle d'Aosta (IT)	−16

All we have done is differentiate the header row, rearrange the order of the data, present it in descending order and delete the superfluous grid lines. It makes it much easier to see what is happening.

Comment	• Many companies have established ways of working. They may not be the most effective but they most certainly will be deeply embedded in the culture of the workplace. In circumstances like this we suggest that you tread with caution. There is also the issue of established practice. In the table above we used the negative sign to indicate a negative movement, yet in accounting it is traditional to use brackets – (£1000) indicates a negative quantity or a loss.

Data and information

It is important to be clear on the distinction between these two terms. *Data* refers to unprocessed information – it has no sense or meaning. *Information* is data that has been processed and is *meaningful* to the end user. If it is not meaningful to someone then it is worthless. There is a case on record in which a worker in a New York stockbroker company, during a ticker tape parade, threw out a lot of old papers as a contribution to the parade. This turned out to be the database and client details of the company and led to other stockbrokers scurrying down the street trading sheets of paper at vastly inflated prices. The point this episode makes is that the data was not meaningful to the person who threw it out of the window, but to those in the know it was priceless.

Communicating through graphs and charts

There is an old saying that 'a picture saves a thousand words' and it is certainly true in the field of data representation. If you draw graphs and charts of your data then the patterns in the data will be quite obvious to all who need to see them. This can affect the judgements of the strategy makers and so it is vital that the graphs and charts accurately represent what the data tells us.

We have already mentioned the fact that politicians, in the UK at least, should not lie directly but they most certainly do use data to distort the picture.

Look at these two graphs.

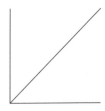 

They seem to tell different stories but in fact they do not. Most people would feel that the performance indicated by the left-hand graph is more impressive than that indicated by the right-hand graph. What's missing are the scales on the axes – there is no information about amounts, times whatever. Without this information the graphs are, at best, misleading and, at worst, useless.

Journal	• Discuss the valid uses of data and the dangers of treating it carelessly.

9

Managing Data

'Whatever you do, do cautiously and look to the end'
Translation of a Latin saying

One-minute overview

Different types of numbers have different functions. A *nominal* number is used to give a unique identifier to a property of something we are keeping data on. For instance, if you give each member of a club a membership number the purpose of that number is solely identification. You cannot add all the membership numbers and get a meaningful answer. *Ordinal* numbers are used to generate order and comprehension in data. For instance, you may need to rank preferences in a questionnaire concerned with opinions. *Interval* numbers are used to represent quantities and we can perform mathematical operations on them.

In this chapter we will look at:

- what is meant by the term 'number';
- issues surrounding the use of computers to manage data;
- maintaining a database;
- cells, records and variables.

Nominal numbers

We usually take the concept of a number in our stride. We take it for granted that there is a common understanding and an effective common 'currency' for numbers but there are exceptions to this, as we shall see.

We have already defined nominal numbers. Consider where they are used in your daily life:

- how many clubs or societies do you belong to?
- what about your driving licence or library card?
- etc.

The useful thing about numbers, and it is not stating the absurd, is that we never run out of them. This gives us an infinite system of representation. This is their value in terms of statistics – they are a functional utility that can be used in many situations.

The majority of organisations now use statistical software to maintain records and numbers in this form, rather than letters or a combination of letters and numbers (called strings), are easier to manage.

Dichotomous coding

We will look at questionnaire design in a later chapter but for now it is important to recognise that many questions have only two possible answers and that these answers are mutually exclusive. By 'mutually exclusive' we mean that the presence or action of one excludes the presence or action of the other – for instance, a person must either be male or female. While there are a small number of people who may have to be described as hermaphrodite, it is almost always the case that a person is either female or male. This type of question is described as dichotomous and are usually coded Y/N or 0/1 or ½ in the data collection world.

Ordinal numbers

Ordinal numbers are used to generate order – rank order. If, for instance, a person is asked to rank the soccer teams in a given list of Arsenal, Chelsea, Liverpool, Manchester United and Wrexham in terms of favouritism, they may respond with:

1   Manchester United
2   Liverpool
3   Wrexham
4   Chelsea
5   Arsenal

(You will notice a bias towards the north west of England and north Wales – presumably where this person comes from or grew up.)

The order in which the teams, or opinions, are written becomes the code order for this question or category. It is clear that the code numbers are not fixed – in our example Manchester United is coded 1 because *that* was the choice of the respondent. However, it is clear that Manchester United could have been coded 1, 2, 3, 4 or 5 depending on the individual's choice.

You need to be clear that ordinal numbers give order to data – they *do not* represent quantity. For example, from the data above it is invalid to say that the respondent liked Manchester United twice as much as Liverpool simply because Liverpool is listed second with a code number a double multiple of that used for Manchester United.

Interval numbers

Interval numbers are used to represent quantities and we can operate on them mathematically and manipulate them. A good example is afforded by how questionnaires ask for information about age. This is a delicate matter and can cause offence, therefore many questionnaires ask for age to be recorded through a grouping technique. For example,

under 18   18–25   26–35   36–45   46–55   over 55

Here you cannot be sure of someone's actual age. You simply know that the respondent is somewhere in the class interval. This allows you to operate on the data you gather. You might be able to say, for instance, that there are three times as many respondents in their 30s than those in their 20s.

Journal	• Make an entry showing your understanding of the three different types of number we have discussed.

Using computers to manage data

Data collection can generate thousands of responses and so it is essential to use suitable software to handle the information collected.

Data collection and entry

When it comes to choosing software you pay your money and take your choice, but there is some sense in using the same software that other people use. In the *Mr Educator* office and at *Studymates* we use *Microsoft Office*. This is not intended as an endorsement of Microsoft but to simply state the fact that it is the market leader. When data is being shared or transferred electronically, common software makes the process much more reliable.

The *MS Office* suite has a number of tools but two of them are particularly useful in handling data – the *Access* database and the *Excel* spreadsheet. This book cannot give a detailed account of their use but there are many books available which can provide that information, at whatever level you need.

MS Access is a powerful tool but it is quite straightforward to use at a basic level. Designing a database in Access is relatively straightforward – if you are new to the database idea then it has a 'wizard' to guide you. One good tip to bear in mind from the start, which will reduce errors later, is to specify what may or may not be accepted in the data fields you use. For instance, if the coding used for a particular item in a questionnaire will only allow a response of 1, 2, 3 or 4 then an entry of 5 should be rejected automatically. This is a good way of checking the accuracy of the data you are entering. Caution is still needed though – it will not help if the response supplied is 2 and you enter 3 by accident. Some schools recommend double-entering data to check its accuracy. However, this takes twice the time and anyway if the data do not match it means that one of the data inputs was wrong – but which?

MS Excel is also a powerful tool and a very functional piece of kit. The spreadsheet can accept data directly into cells but this can be 'automated' through the use of an Excel form.

The choice of use does not really come down to one or the other – it is a more a case of six and two threes. Excel is not really as user-friendly as Access when it comes to data entry and reporting but it most certainly does have the edge when it comes to data manipulation. It is possible to set up the data in Access, export it to Excel for data manipulation and then back to Access for reporting.

Cleaning your database

With the best will in the world, it is inconceivable that there will never be errors made during data input. Therefore we advise you to adopt a policy of checking the accuracy of what you have entered as soon as you have finished a section or a session. This can be extremely tedious with large datasets but some strategies make it easier to manage.

- If the data set is reasonably small, then print it and check each entry individually against the questionnaire responses.

- If the dataset is large, then print a table listing each of the variables in your dataset. Don't forget to include the frequency of each code – this should show up those codes that should not be there.

The essential point about collecting data is that you need to have an accurate data to input. Collecting accurate data depends on asking the right questions and that, in turn, depends on using the right questionnaire – the topic of the next chapter.

10

Questionnaire Design

'There is no such whetstone to sharpen a good wit and encourage a will to learning as is praise'
Roger Ascham

One-minute overview

Most data used in statistical analysis is collected through questionnaires. In a job situation, you may be asked to be involved in market research or focus group management. This is the ideal situation to use a questionnaire – but it is also where huge and costly mistakes can be made. We mentioned earlier a local council which noted a huge (95%) positive response to a questionnaire they distributed to determine the demand for a games area in a village. They went ahead at significant cost only to find that the facility was under-used. Unfortunately they had failed to notice that the 95% majority was from only a 10% return. Their pitfall was the design of the questionnaire. It contained inappropriate wording, which caused many to not bother replying, and *led* those who did respond towards the answer that was required.

In addition to poor wording, the question order can help motivate respondents to answer – however, it may also confuse them. If you are constructing a survey, then you should first of all pilot the draft questionnaire and amend it accordingly. It can then be despatched with a covering letter explaining the purpose of the questionnaire along with *clear instructions* for the respondent.

In this chapter we will look at:

- the reliability and validity of data;
- the structure of the questions;
- the structure of the questionnaire;
- the pitfalls to avoid in data collection;
- the pilot survey;
- the covering letter.

Reliability and validity of data

Think about this question – how aware are you of your own bias? Whenever you collect data you must ensure that the whole process is as free from bias as possible. Your data collection needs to be both reliable and valid.

Reliability

The only way to judge if the data collected, and by implication the questionnaire, is reliable is to have a measure of the consistency of the data responses. By this we mean that the responses to a questionnaire given to a random sample of individuals should be compared to those from another randomly selected group given the same questionnaire. If the pattern of response is similar then the questionnaire is generating reliable data. Reliability happens when the data can be repeated – we call this test–retest reliability.

Validity

By validity we mean the effectiveness of the questionnaire in gathering information about the items mentioned in that questionnaire. This depends on how well you have constructed the questions and the questionnaire overall. The respondents must understand what questions you are asking and you need to be clear that there are no ambiguities in the wording of the questions or the structure of your questionnaire.

Although you may design a questionnaire that is reliable, it can still fail if it does not measure the concept you wanted to target. Reliability is not the same as validity – this is why the pilot stage, so often overlooked, is an essential part of the process.

The structure of the questions

The type of response you require from each question can affect the way a respondent answers. If the respondent has difficulty with the question, or finds it demeaning, there may be no response. There may also be sarcastic or inappropriate comments written on the response mechanism! This may happen anyway but the best strategy is to make the questionnaire as clear as possible.

Open or closed questions?

Open questions are those to which a response other than 'yes' or 'no' is required. For instance, 'Where do you live?' is an open question – the respondent cannot, sensibly, answer 'yes' or 'no'. Similarly 'What type of music do you listen to?' is open.

The question 'Do you work?' is a closed question as it can only be responded to by 'yes' or 'no' or 'sometimes'. In data collection, closed questions provide data that is easily coded for entry into a database.

Closed questions

There follows a typical closed question.

From the list below, tick the one item that best describes your pattern of listening to music.

I enjoy a variety of music and get bored with one type		(1)
I think classical music is the most relaxing, it calms me down		(2)
I listen to my favourite radio station only, it is important to listen to new music		(3)
Other		(99)

Each option has been given a code in order to make the data entry task easier. We have included 'other' as a category because this can be a useful device. Some respondents may not recognise their listening pattern in the options list but they still need to make a response.

By coding 'other' as 99, it gives you the opportunity to include new codes, from 4 to 98, should you need to do so in future. One way of doing this is to follow this closed question with an open one:

If you ticked 'Other' please describe your listening pattern here

You can then analyse the data collected and determine an improved set of options to include in future surveys.

Open questions

There are times when you will not want to impose categories of response on the respondents. It is also a useful device in enabling categories of response where you have no clear idea as to what they may be.

Here is a good example:

In the space below give one clear reason why you want to join this company.

This gives respondents greater control and often encourages them to convey how they really feel. This technique has been used successfully in business where people use laptops to respond to questions anonymously. It has been found is that when people feel free of restraint they are more likely to say how they really feel, rather than how they are *expected* to respond. By common agreement, this type of question allows respondents to give richer, more informative, honest answers that genuinely draw on their experience and understanding. However it can also be a real headache for the data processor. How can you code such potentially wide-ranging responses?

Journal	• What is the basic difficulty in this coding problem? Are there any solutions?

One approach is to anticipate at common areas of response and then generate coded categories that are likely to cover all (or most) responses.

On important point to remember is that if you use open response questions then you are likely to be creating a lot of work later on when processing the information.

Single or multiple response questions?

The examples given above are called 'single response' questions. This is because the respondent can only answer by choosing one item. Multiple response questions are those in which respondents are asked to provide more than one response to the question asked.

The usual method of doing this is to use scales, ratings and rankings in questions that require a response to every item. This restriction of choice means that you are forcing the respondent to identity a specific number of reasons for thinking, doing or preferring something and this will add value to your data collection.

An alternative approach is to provide a set of possible answers and to limit the respondent to a *maximum* of, say, three or four. This means that a range of circumstances are being catered for and that if some of the options do not apply to them personally, they are not obliged to 'tick' a specific number. In other words, they can indicate with fewer responses. It also means that you, as the collator of the information, don't have to manage a mountain of responses.

This is something that we have to stress. Whenever you collect data you must have a clear idea of how you intend to process it. We know of one school in the South of England where the headteacher went ahead and spent a lot of money sending out a questionnaire to parents. There was a good return, over 80%, but they had no idea about how to process the data. It took something like 15 working days to process the information, and even then it told them very little. This is a good example of poor planning.

The structure of the questionnaire

The ordering of questions and the overall look of your questionnaire are vitally important. In developing the questionnaire you must try to make sure that respondents are motivated to finish *and* providing the required answers. This is especially true with a long questionnaire. The purpose should be to create a document with progressively graded questions that lead the respondent through the process. The structure of your questionnaire will affect its reliability and validity.

Funnel ordering
In this approach, respondents interpret your questions in their own terms and within their own context. The order of questions can send unintentional messages about perceived intentions. By this we mean that the response to one question can affect the nature of subsequent responses.

This type of questionnaire begins with questions that are general and broadly-based but progressively become more narrowly focused. This approach works when:

- your respondents are well motivated;

- you need to find specific detailed information (the funnel approach is particularly good for this);

- a particular question may impose a frame of reference before responses are given, and so influence the responses.

Here is an example of the funnel approach – the questions are not written in a questionnaire-friendly way but do illustrate the point.

1   *List some of the ways in which the UK benefits by having the 'special relationship' with the USA.*

2   *Which one of these benefits is the most important in your opinion?*

3   *Why do you believe that this is the most important?*

4   *Where do you find your information about this benefit?*

5   *Which newspaper(s) do you read?*

It is clear here that the overt agenda is to see how newspapers affect opinion on political matters. By funnelling the data we are forcing the respondent to think about the relationship between the UK and the USA. Then, at the end of the funnel, we determine what we want to know – namely which newspaper do they read. The purpose is clearly to see how perceptions are shaped by the newspaper they read. The 'which newspaper' question comes last deliberately – if it was listed earlier it could bias subsequent questions.

Journal	• Make up of another example of funnel questioning.

Filter questioning

There will probably be some areas of your questionnaire in which the questions apply to some respondents but not to others. This means you may need to divert people away from irrelevant questions to more appropriate areas.

Here is an example.

1 *Do you have children?*
   If no go to question 2; if yes go straight to Question 4.

2 *How many other people live in the same home as you?*

3 *Do you have security features in your home?*

4 *What are the age(s) of your child/children?*

The response to question 1 filters by directing those respondents with children to question 4 immediately, precluding them from irrelevant. This approach also helps to ensure that all respondents will be more likely to complete the questionnaire. It is a useful technique to bear in mind when designing some areas of some questionnaires.

The pitfalls to avoid in data collection

From personal experience you will probably be aware that many gaffs are made in questionnaire design. While some respondents may smile at a double entendre, however innocent, others will find it irksome – it also makes you look unprofessional and casts doubt on the integrity of the questionnaire as a whole.

Language and wording

You are strongly advised to keep the language used as simple as possible. Lengthy questions can confuse and lead to misunderstanding that has a dispiriting effect on respondents There may well be occasions where explanations are required but use these sparingly. Such explanations may be more appropriate if the response involves choice – as in multiple-choice items. This will mean that the length of the explanation preceding the question is counterbalanced by the shorter time required to answer it.

Also try to make sure that the language used is appropriate to the audience. Jargon is context-specific language, which is best avoided unless it is essential or appropriate.

Leading questions

Such an approach often guarantees that many respondents will give the answer it directs them to give. It is natural to try to be 'helpful' when designing a questionnaire. There is often no deliberate attempt to mislead – just a desire to be user-friendly. In the local council survey mentioned earlier, we spoke of how the council had created a biased survey. One of the questions asked 'Do you agree that it is a good idea to have sport facilities for the young?' It is hardly surprising that they received an overwhelmingly positive response. The question led the respondent to providing the answer that was required. Most people would agree that it is desirable for young people to engage in sport.

It would have been far better to start the questionnaire by asking the respondents if they took part in sport themselves and/or if they had children. This would probably have given a better picture as to the thinking of respondents.

Be aware that there is an area of study known as 'expectation theory'. People will often not answer as they truly feel but in the way they feel is expected. For instance, smacking children is an issue in the UK. If you were to ask people if they agree with such a sanction for children who have misbehaved, many may say they do not agree with it and yet do smack their own children. This is an example of giving the answer that they feel is expected and not the one that represents their actual opinion.

Questions that threaten self-esteem

These are questions that make the respondent question their values, beliefs and behaviour. If you ask questions which imply that the respondent is not acting in a socially responsible manner, they will probably be badly received.

Such questions threaten the self-esteem of the respondent.

Embarrassing questions may also be potentially threatening and can be taken as an invasion of privacy, or even ridicule.

Journal	• Write down an example of a question which might cause offence, but in a subtle and not blatant way.

Unintentional multiple questions

These are questions that the respondent may not know how to respond to because more than one response is possible. For instance:

'There is far too much sex and violence on television these days'. Do you

strongly agree   agree   not sure   disagree   strongly disagree?

The respondent may feel strongly about sex on television but not violence and is therefore unsure how to answer the question.

A simple solution here is to either split the item into two questions or have two separate scales – one for 'sex on TV and the other for 'violence on TV.

Ordering the questions

Using funnel sequencing will help in developing order in some items in your questionnaire. However, your questionnaire could be complex overall covering a number of areas using a range of questioning techniques. If this is the case, then you need to structure your questionnaire very carefully.

In general:

- ask the straightforward questions first;
- it is often better to place closed questions before the open questions;

- if you have questions that are complex, or relate to sensitive issues, you should make sure they appear late in the questionnaire.

This strategy will allow your respondents to feel comfortable with taking part, before confronting more challenging questions. Even if a respondent refuses to continue past a certain point you will have gathered some useful data.

The covering letter

If you are gathering data by post, the questionnaire will need a covering letter. The letter must explain the purpose of the questionnaire and how the information will be used. Most people will be indifferent to you unless they have a vested interest in the survey, so it will be to your benefit to convey the value of the survey to the respondent early on – especially if they may benefit from the results in some way.

Your covering letter should cover the following areas.

- Identification of the person or organisation carrying out the survey.

- A description of the nature and purpose of the survey.

- A statement about why it is important for the reader to respond (the benefit to the reader). Questionnaires often carry an accolade – for instance, 'We are looking for the sort of person who ...' – followed by something positive. The danger is that this can bias what comes next in the survey so be careful when using this approach.

- Confirming the recognition that all information is confidential and that there is no real or implied threat in respect of harm or embarrassment.

- A statement that the requirements of the Data Protection Act have been satisfied and that the data supplied will not identify any individual or organisation.

The pilot survey

The pilot survey is a small-scale version of the actual survey you intend to undertake. Its purpose is to iron out any problems concerning the nature and structure of the questions. Listen carefully to the feedback you are given and make appropriate amendments, but be careful not to build in untested problems!

One way of doing this is to send the questionnaire and covering letter to a small sample of people and check their responses. This affords a way of judging the performance of both the individual items and to the overall response to the questionnaire. Some regard this as an unnecessary step but it is crucial. It is a cheap way of testing the survey before you incur the full expense.

If the pilot is carried out properly it will provide you with an opportunity to modify questions, to eliminate ambiguity and to tidy up the questionnaire generally. You must make sure that the pilot is carried out under the same conditions as you intend the real survey to be, but make sure that the data received from participants in the pilot sample are excluded from the actual survey.

The size and extent of the pilot survey will depend on the nature of the survey as a whole but it is vital that the pilot sample is chosen using the same random process that you will use in the actual survey.

| Journal | • Summarise the main stages in constructin a survey to be carried out through a questionnaire. |

The next stage is to summarise the data you collect and that is the topic of the next chapter.

11

Handling Data

'*A wise man will make more opportunities than he finds*'

Francis Bacon

One-minute overview

You have undoubtedly heard the famous quote 'lies, damned lies and statistics' which many people wrongly interpret to mean 'statistics tell lies'. Statistics are simply inanimate measures, or numbers, that indicate patterns – clearly they themselves cannot lie. But it *is* true that liars often use statistics to justify their viewpoint. Therefore you do need to develop an intuitive confidence when dealing with data. This will happen when you understand the basic underlying concepts.

In this chapter will look at:

■ different types of average – the mean, the median and the mode;

■ identifying discrete and continuous data;

■ displaying data in charts;

■ using scattergraphs to show correlation between variables.

Importance of data handling

Data handling is part of every job. It does not matter if it is a high-level post with a big company or a self-employed tradesperson, there will still be a need to handle data. One common word that you need to understand is 'frequency'. In data handling this refers to the number of times an event occurs.

Something else to beware of is that data presented in the media may be misleading. Politicians often present data in a manner designed to influence how the public thinks about the issue they are proposing or defending. Those who do not understand the nature of the data can easily be misled –and this can cost you dearly.

The three averages

You will no doubt have read many times about the 'average man and woman'. This is a notional couple who are used to demonstrate the effect of government policies on the nation as a whole. But what is an average?

In fact there are three types of average – the mean, the median and the mode.

The mean

This is the measure worked out when all the data are added together and the resulting answer is divided by however many items of data there are. The easiest way to understand this is to look at an example. Let's say that three people earn the following amounts weekly:

£300   £120   £200

To find the mean of these three numbers, we add them and then divide by three (because there are three items of data). Now £300 + £120 + £200 = £620 and £620 ÷ 3 = £206·67.

So the mean wage is £206·67 (to the nearest 1p). In effect, what we have done is to notionally distribute the money evenly between three people. In reality none of these people actually earns this amount but it does give us a measure against which to judge how well they are doing – the person earning £200 per week comes closest but the person earning £300 per week is well above the mean. The point to remember is that a mean figure is not necessarily part of the original set of data.

| Journal | • Is this always true? Make an entry investigating this. |

Here are some examples to make sure that you are familiar with the idea of the mean.

Example 1: Find the mean of the data set 1, 2, 3, 4, 5, 6, 7, 8, 9

Sum = 1 + 2 + 3 + 4 + 5 + 6 + 7 + 8 + 9 = 45

There are nine items altogether so the mean = 45 ÷ 9 = 5.

One way to understand this is to think in terms of real things. Imagine that 1 to 9 represent nine different children. Child 1 had 1 sweet, child 2 had 2 sweets and so on. For the mean number of sweets just ask 'How many sweets would each child have if they were evenly shared out?'

Example 2: Find the mean of the following amounts of money.

£3·20 £35·30 £19·00 £2·50

Sum = £3·20 + £35·30 + £19·00 + £2·50 = £60·00

There are four items of data so the mean = £60·00 ÷ 4 = £15·00.

Example 3: Find the mean of this set of data.

2   5   7   3   8   9   6   4

Total = 44. Number of items = 8. Mean = 44 ÷ 8 = 5·5.

The median

The median is the middle item in a set of data when the data are arranged in order of size, smallest to largest or largest to smallest. The following examples illustrate the idea.

Example 1: Finding the median with an odd number of items.

4   5   8   9   7   3   6   2

We must first rearrange the items of data in order of size – we will arrange them in order of increasing size, but the other way would give the same answer:

1   2   3   4   5   6   7   8   9

Therefore the median must be 5, since this is the middle term (there are four numbers on either side).

Example 2: Finding the median with an even number of items.

1   3   4   2

There is no middle item to pick so we need a slightly different approach.

You still need to put the data in order, either ascending or descending.

1   2   3   4

Now we take the middle two terms, 2 and 3, and work out the mean of these two numbers.

Sum = 2 + 3 = 5. Mean = $5 \div 2 = 2 \cdot 5$.

So the median of this set of data is 2·5 even though 2·5 was not actually in the original set of data.

The mode

This is the item of data that occurs the most often in a particular set of data.

Example 1: Find the mode of the following set of data.

1	2	3	5	7	2	8	5	9	1	5	176
189	1	234	1	66	55	7	33	48	39	1	2
55	7	15	12	1	4	5	6				

To work out the answer to a question like this, we draw up a *frequency table* by using the tally method.

Item	Tally	Frequency
1	┼┼┼┼ │	6
2	│ │	2
3	│	1
4	│	1
5	│ │ │ │	4
6	│	1
7	│ │ │	3
8	│	1
9	│	1
12	│	1
15	│	1
33	│ │	2
39	│	1
48	│	1
55	│ │	2
66	│	1
176	│	1
189	│	1
234	│	1

From this frequency table, it is easy to see that the number 1 occurs the most often – therefore 1 is the mode.

Example 2: Find the mode of the following:

Blue, blue, red, blue, white, white, green, white, red, white, white, puce, white, pink, white.

Sometimes you can just scan the data – in this case it is easy to see that the mode is the colour white.

The range

This is another useful measure to apply to some sets of data – it indicates the spread of the data. We find this by subtracting the lowest value from the highest value. For example, to calculate the range of

$$12 \qquad 23 \qquad 13 \qquad 46 \qquad 25 \qquad 26 \qquad 17 \qquad 8$$

you should scan the list and pick out the two items needed – here the highest value is 46 and the lowest is 8. So the range = 46 − 8 = 36.

Using averages

Many people think of averages as hypothetical things and as not being of much use but, used properly, they can do quite useful jobs.

The mode (or modal class) is useful in many ways. Obviously there are many situations in which personal provision cannot be made so that every size is catered for – for example, the amount of legroom on buses, trains and aircraft is determined by the population mode. The height of doors is determined by the modal class of population height. You will notice this if you visit many old buildings – you may have often ducked through the doorways of old houses because when they were built the doorways were made high enough for the people of the time. Improvements in diet have lead to the population of today being taller (on average!). (Some US television cop shows used this idea for a 'cheat' during the 1970s – they made the doorways on the set $^{6}/_{7}$ of the normal size. This made the actors look taller relative to the set, and therefore more macho.)

The clothes industry also provides a good example of the use of modal classes. There are bespoke tailors of course, but not everyone can afford to have their wardrobe made-to-measure. On the other hand, bulk manufacturers can't just make clothes for the average (modal) sized man or woman – in the UK the average sized man is about 1·8 metres (about 6 feet) and the average woman is about 1·65 metres (about $5^{1}/_{2}$ feet). Commercially it might seem sensible to make clothes to fit people of this size – but what would most people wear then? Clothes manufacturers have masses of data on body size and distributions that they can use to work out the modes for different size ranges, and knowing how these are distributed across the population enables them to estimate how many items to make for each 'size'. Whether these will sell or not is another matter – but no doubt they have collected statistics about that too.

There is much evidence that the mode, or modal class, is a real-life entity and not simply a mathematically abstract concept. A well-known high street chain in the UK gathered data on women's body size and found that most of them wear the wrong size of bra. By targetting the modal-sized women, rather than the mean-sized woman, the garment is found to fit more women and is reported by more of them to be more comfortable – this increased sales by over £50 000 per week.

A simple example of the effect of understanding how to use the median is afforded by analysis of weight data in the USA. In one state, according to press reports, the median weight for men and women is now about 125 kg (280 lb). This is the *median* weight, so there are many people who are heavier than this. This has a knock-on effect on the provision of goods and services. People this heavy will need bigger cars, larger spaces on aircraft and will cost more to transport. They will also die 'sooner than they should'. There is no cause for complacency here in Europe – we are also getting bigger and many are obese, which means more than 30% overweight. One well-known media personality in Wales told us that he once had a body mass index of 42% – nearly half his body weight was fat. He has subsequently lost a lot of weight, and we congratulate him for this achievement, but the point is that there is an epidemic of obesity and the effect will be dramatic.

Between 2010 and 2025 we can expect to see an increase in youth mortality. Young people, for the first time in over

100 years, will not live as long, on average, as their parents. At the start of the twenty-first century young people are taking little exercise and are smoking more (particularly young women). They are also drinking more (in 2004 UK drink-driving related incidents increased for the first time in ten years) – add to this the use of recreational drugs and you have a cocktail for death. This predicted rise in mortality *will* happen – it is already in the system and, for some reason, has become ingrained in youth culture.

Guns and statistics

Statistically, the most vulnerable group in society are young men. In researching the data for this book we met a carer from Manchester who is mentoring young people who belong to gangs. Gang membership is on the rise – following the US model. It gives young men an identity, something to belong to that generates a feeling of esteem or well-being about themselves. The bizarre fact is that 13 year old children in Manchester (and other cities) are literally 'ducking and diving' through inner city gardens and back streets dodging bullets – yet another a cause of death amongst the young, and it is often the innocent who suffer. There have been high profile drive-by shootings in Birmingham, where two young women were killed, and in Nottingham, where a teenager was murdered as she returned home from a funfair.

Why mention this? To encourage parents to share these facts with their children. It is never too early to give your child responsibility. Make sure they understand the concept of 'the average' and encourage them to be 'above it' in matters of health and behaviour. In simplistic terms, allow them to eat chocolate but limit it to one small bar per day – on the day of writing we saw a girl of about twelve stuffing a huge bar of chocolate down at 7.45 am, hardly a healthy way to eat.

Politics and television

At general election time in the UK, or presidential election time in the USA, the phrases 'the average wage' and 'the average person' are used rather freely. As a statistically enlightened voter you may now be a little sceptical when you hear them because it won't be clear what the speaker means by them. Are the phrases being used in the 'everyman everyday terminology' context or is the speaker referring to the mean of the set of wages, the mean of the set of people? Or the mode or the median? A healthy dose of cynicism is useful when dealing with and attempting to understand data when politicians, or the media, present it.

Journal	• Make an entry about the relationship between statistics and cynicism.

Part Three
Money and
Finance Matters

12

Managing Your Money

> *'Give us the tools and we will finish the job'*
> Winston Churchill

One-minute overview

Have you ever had the thought 'where has all my money gone?' Are the children constantly asking for more when there is little left in your purse or wallet? This is where this book can help.

In this chapter we will look at:

- techniques to help improve your money management;
- the psychology of debt;
- planning tools that will enable you to plan ahead.

The top idea to get to grips with is 'cash flow'. This is just what it says – the flow of cash into and out of your life. The trick is to have more flowing in than flowing out, and it really isn't all that difficult to manage your family's spending patterns, so long as you want to. The challenge for many is to move away from the thinking that suggests there is a shortage of money to the type that suggests there is an abundance of it.

Every day $1·2 trillion circles the globe looking for opportunities. This is what globalisation means – money is electronic, it is looking for a home and, if you understand the system, that home could easily be your pocket.

Do you really want to be debt free?

Who doesn't? But this is not such a crazy question as it seems. Throughout this book we have asked you to keep a journal, a notebook that examines and records what you are thinking at various stages in your mathematical development.

Journal	
	• Be honest – do you really want to be debt free? Do you want to 'pay the price' for this because to be debt free you will have to change your thinking and spending habits. • Make a positive statement, an affirmation, now. Make it your aim, with the help of this book, to get spending under control. Write it down now – 'I will control my money and make it work for me' is a good affirmation. • By writing it in your journal and dating it, you are making a commitment to take control of an important area of your life and are looking to improve the future for yourself and the ones you love.

'Believe in yourself, you do have talent'
Graham Lawler (aka *Mr Educator*)

Journal	
	• Write this down. 'In five years time, I will be a [*put your job here*]. This will give me [*put your wages here*]. I will save 10% every week, this will be [*put the amount here*].'

Congratulations! You have taken a major step forward – but now you need to decide where your money goes.

Look at this table – it shows the spending pattern of a teacher and his family. Teachers get paid monthly so we have set this up for the calendar month – if you get paid weekly, just do the same on a weekly basis. To keep thing simple we quote this teacher's net pay – that's the pay received after tax and national insurance contributions have been paid.

Fixed costs	Amount	Amount remaining
Monthly salary	£1500	£1500
Mortgage	£750	£750
Food	£360	£390
Entertainment	£300	£90
Travel	£150	−£60

Let's stop here – straightaway we can see trouble. There is simply not enough money to go round, or is there?

Journal	• Do the same exercise for you and your family. If you feel happier doing it on a weekly basis that's fine, but please do it now before moving on.

In the teacher example we looked at the expenditure of the family and it was quite clear they were spending far too much. We left that exercise with a question hanging – is there enough money?

To be honest, for some people there would never be enough money – this family may be one of them. In surveys many people say that if they only earned an extra 25% then everything would be fine, but the truth is there would never be enough. If they did get another 25% they would only spend it and would still be in debt. The fact is that, for some families, spending has got out of control.

Let's examine the teacher's family spending more closely. Each month they spend £360 on food. They argue that they always shop in the cheapest supermarket and simply do not understand where the money goes. Yet they fail to reveal that they also take the children to a burger bar on a number of occasions each month and the cost of this works out at £60 per month. That's a whopping £720 ($1500 or €1000) per year on burgers – junk food to boot. Note that they are £60 short each month relative to their income, so if they stopped taking the children to the burger bar the deficit disappears – *and* some surplus appears!

Further investigation of their spending patterns shows that there is scope for savings in the entertainment budget. On examination, the teacher admits to going out three nights per week with his friends, leaving his partner at home with the children. On average he spends £15 on each night out – that's £45 per week, £180 per month, on entertainment. The teacher has to question if this justified.

In another analysis, a publishing executive and her family found that they were spending £1500 per year on magazines. The family cut this expense down drastically and found that were able to afford a holiday that they previously felt unable to – and it didn't involve any saving-up.

> *'They can because they think they can'*
> Virgil

We also found a third example concerning a family in which the father was an avid music fan and could always find money to buy discs. He had a £7000 music unit and something in the region of £5000 worth of CDs, and yet he complained that he could never afford to take his family abroad for a holiday. The sad fact is that he has failed to recognise where his money is going. He buys music on impulse and can always justify the expense – £20 here, £15 there and so it goes on. He and his family live on credit simply because he has not taken control of his spending habits.

In another example we looked into the costs of a smoker. This young woman smokes 20 cigarettes per day. Let us, conservatively, assume that these cost £4 per packet. That means she spends $7 \times £4 = £28$ per week, or £1456 a year. This is money which could be more usefully directed – *and* she would be far healthier.

How in control of your spending are you?

We credit the American writer John Cummuta for his *Debt to Wealth* programme for this idea. Cummuta pioneered an approach to wealth creation and has used it himself. He was formerly a top executive with a top executive lifestyle earning around $350 000 a year. All was well until he lost his job when the company downsized. He had been running a gold-plated car – people would be impressed and say 'Ooh what a lovely car!' Only after he lost his job and had to examine his lifestyle did he realise that he had been paying $50 000 a year simply to hear other people say 'Ooh what a lovely car'. How much are you paying just for the privilege of hearing other people's comments?

If you have to have to buy a flash car on hire purchase do you realise what 'APR' means? We will look at this in the Chapter 13, but for now let's suggest that if you are planning to buy the car then why not look for a cheaper option – or do other people's 'oohs' matter a lot to you? The truth is that many of the rich do not drive expensive cars. Many millionaires drive ordinary cars – Fords, maybe Toyotas – and they buy their clothes at standard high street stores and live within their means generally. Yes, even millionaires set budgets *and stick to them.* They know the difference between assets and liabilities do you?

Journal	• Write down what you think are assets and liabilities. What about *your* assets and liabilities?

What are assets and what are liabilities?

Put simply, as we said in Chapter 6, US writer Robert Kiyosaki writes 'assets feed you, liabilities eat you'. In other words, assets bring money in, liabilities leak money out. Another way of putting it is that an asset is something which creates a positive cash flow while a liability creates a negative cash flow.

A car costs you money, so it is a liability – it needs to be taxed, insured, fuelled and serviced. If you have a dog it is a liability – many have to be bought initially and they all have to be fed; some get injured or fall ill and vets are not cheap; some have to be kennelled while you are away on holiday. But some people offset this liability by breeding dogs, which they sell on. It is possible to turn some liabilities into assets. I'm not saying you should, but your car could be used as a carrier service.

'Nothing has power over me, other than that which I give it through my conscious thoughts'
Anthony Robbins

Getting out of debt

The first step is to analyse your spending patterns. Look for anything that can be cut – be ruthless and cut it out. For example, look at ways of making your own sandwiches for lunch, for a fraction of the price, rather than buying them every day over the counter. Look at the figures, in London there are sandwiches on sale for £3·75 – multiply that by 5 (days a week) and then by 48 (working weeks a year) and that is the annual cost of these sandwiches. By getting up 10 minutes earlier to make your own sandwiches, and shopping at your local superstore, you can save well over £700 a year – just by making your own sandwiches you could have an extra £7000 spending power in ten years. Or you could use it to make a regular investment and make even more money. Can you see how powerful these cost saving measures can be?

Look at your spending patterns and try cutting back on whatever it is you spend – drinks, magazines, theatre trips etc. Being an adult involve being responsible – if you and/or your partner are finding this difficult, you need to look carefully at your lifestyle.

We know of one couple who went through this analysis and transformed a debt, which ran into thousands of pounds, into savings within three years. They bought houses and then rented them out. The rent which came in was more than the cost of the mortgage they had had to take out so there was a small amount left over – they had a positive cash flow. They now own several properties and are slowly building up their wealth. Meanwhile, the value of the houses they have bought is increasing – partly because of inflation and appreciation, but also because of the improvements they have made. They cannot believe how much their lives have been transformed – both still work hard but they now concentrate on buying assets rather than incurring liabilities. In just three years of disciplined investment they turned a debt into wealth.

The secret

Direct your expenditure away from buying liabilities towards buying assets. Investing in your, or your children's, education counts as buying an asset – it empowers the making of money in the future.

This is how rich behave – rich people acquire assets to generate positive cash flow. They minimise liability purchasing to reduce negative cash flow to a minimum.

Consider this – if you won the lottery and decided to buy a luxury yacht would you:

- spend all your time on the new yacht?

- spend some time on the yacht and leave it in the harbour for the rest of the time?

- spend two weeks on the yacht and then rent it out for the rest of the year?

• Be honest with yourself – describe your reaction to each of these options.

There are no right or wrong 'answers' but your reactions will say something about you and your current thinking. Generally speaking, a person with traditional working values will probably say that they want to spend all their time on the yacht. The middle option is typical of the middle classes while the rich would usually give the third response.

Why do rich people, who have lots of money, rent out their yachts? Simple – renting it out generates positive cash flow, so they use the yacht for two weeks per year *for free*. Yes, literally for free – renting out the asset means that they have more money coming in which will enable them to go on and buy further assets. Eventually they may have more money coming in than they actually need to live – they do not have to work unless they choose to. This is quite a goal to aim for –financial freedom, independence and flexibility.

This is the secret of getting out of the rat race – when you have acquired enough assets to bring in more than you need to live, you have left. You are then on the way to becoming wealthy. Notice the subtle message? We are encouraging you to stop working for money and start making money work for you. The first diagram below shows what happens to your cash flow when you simply buy products.

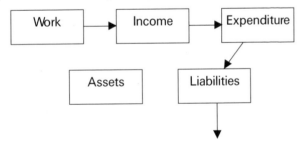

This is probably the spending pattern of most of the poor. Contrast that with the pattern shown below.

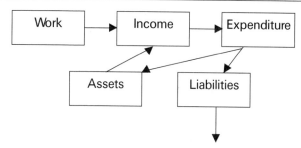

This shows what happens when you buy assets – they generate income. Eventually, when you have enough assets, your flow of income is greater than you need for your living expenses. This is when you become financially independent and you can stop working if you want to.

What about you?

Look at the spending patterns above and compare them to your own. There are some liabilities that we all have to buy – we all need to travel, so that either means paying for fuel or for passenger ticket; we all need homes to live in, fuel to warm them, clothes to wear, food to eat. These all take cash away from you. The trick is to minimise this while still enjoying a good quality of life within your means. In the 1980s we had the rise of the 'yuppie' and we started living on credit – a practice in which we simply spend next month's salary before we receive it. Yet the parents of this generation were brought up differently. The parent generation were used to living within their means and that is something that we need to get back to. So the first thing is to live within your means, the second is to save. A reasonable aim to have is to save a minimum of 10% of your income. These savings will become the means to purchase assets.

Brand name goods

In late 2004 the BBC showed a documentary about the power of brand names and their effect on children. A nine year old girl was shown in her various outfits. Her mother spent literally everything she earned on her children and their clothing, and they had to be brand named. The girl had a term for non-branded goods – 'nicky no names'. These were to be avoided at all costs. That meant that mum was saving *nothing*. She was spending everything she earned on liabilities and will simply not be able to stop working – her job had a special significance, 'just over broke'.

Are these trivial arguments?

If you think this is trivial then think again. The American author Robert Kiyosaki notes in his books (the *Rich Dad, Poor Dad* series) that some 77 million Americans are coming up to retirement in the next ten to fifteen years. What Kiyosaki predicts for the US is also true for the UK. By 2015 the baby boomers will be retiring in vast numbers. This means they will want pensions and medical care from the National Health Service. But there will be fewer people in the working generation – there will simply not be enough cash to pay for the retiring baby boomers. As this book is being written we have Labour government but the same would apply if we had a Conservative government – there will have to be an increase in taxation to pay for the increased burden. One possible way forward is to promote an increase in the immigration of skilled workers. In the south of England (in 2005) there is a shortage of labour. There are plenty of jobs but we simply don't have the personnel qualified to do the work – if you doubt this just try finding a dentist with spare capacity. Bringing in immigrant labour will increase the tax yield and increase the amount to spend on benefits and services. However, there is an old saying that 'nothing in life is that easy'. There are political considerations. In 2005 there was no political will to increase the rate of immigration. In fact some political parties wanted a quota on immigration.

What is the downside of doing nothing?

We can only speculate, but we base these comments on what is happening in the US – there are old people in the USA who are too poor to buy food, some living literally on cat food. If we take no action this is a distinct possibility here in the UK. The days of depending on the state have to come to an end. Although we will continue to have care via the National Health Service we simply cannot expect the state to care for all our needs – we need to develop our own knowledge; we need to understand the numbers.

It helps to keep charts to manage the cash flow and that is the next topic of the next chapter.

13 Cash Flow Forecasting – Profit and Loss

'Things do not change, we change.'
Henry David Thoreau

One-minute overview

Managing money in your life means using tools that can help you keep track of your spending.

In this chapter we will look at:

- cash analysis;
- cash flow forecasting.

Cash analysis

Put simply, this is examining where your money goes. It is vital for your long-term welfare that you can determine exactly where the money is spent.

You should identify the main areas where you spend money in your household budget. If you are not sure how to do this, keep a diary and write down everything you buy – and we mean *everything*. These are your cost centres. Cost centres are split into two types – fixed costs and variable costs.

Fixed costs are those that you don't have to estimate – you know exactly how much you are going to spend on those items. For instance, if you pay rent for your living accommodation then you will know how much this costs – it may go up every six months or every year but at any particular time you can treat it as though it is a permanent fixed cost. Let's assume your fixed costs are:

- rent/mortgage
- council tax/local tax

Variable costs are those that can fluctuate. Typical variable costs are:

- food
- utilities (power/ water)
- travel
- entertainment
- savings
- childcare.

Other costs will depend on your situation.

The next step is to make a concrete record of your spending (see page 136).

Journal	• Set up a table like the one on page 115. Here we have set up the cash analysis on a weekly basis for a person living in the UK. Some sample figures have been entered for the sake of the discussion which follows. You put your own spending figures in your table.

This is called double-entry bookkeeping. We have entered the costs centres under the column headed *item*. We have put the amount spent for each item in the *spend* column and also under the headings that are most appropriate. Notice that there are two entries relating to travel. The first we can regard as the normal monthly expenditure for travel to work. The second is the added expense of a one-off train journey. In the bottom row and last column we have totalled each cost centre – this gives a double check.

This technique allows you to see exactly where your money is being spent. This is the first step to managing your

Item	Housing	Council tax	Food	Utilities	Entertainment	Savings	Travel	Childcare	Spend
Rent	£50								£50
Council tax		£12							£12
Food			£55						£55
Utilities				£15					£15
Entertainment					£30				£30
Savings									
Travel							£30		£30
Train							£35		£35
Childcare								£50	£50
Total	£50	£12	£55	£15	£30		£65	£50	£277

money. The weakness in the hypothetical spending pattern in the table is that there is no record of savings made It is vital that you apply the Cummuta principle we wrote about earlier.

Just as a reminder, John Cummutta's *Debt to Wealth* plan encourages you to identify areas of expenditure that can be converted to savings – this is the key to your success. If you can identify expenditure on liabilities that can be stopped, you can create savings without needing to increase your income. Can you see how powerful this principle is? John Cummuta has given us a tool, or mechanism, with which you can determine your expenditure and identify where you can divert funds from wasted expenditure to savings. If you have trouble saving 10% of your income, try saving 1%, that's just one penny in every pound, then double it to 2% and so on. The target must to get savings up to 10% – this will then build a pot of cash that you can use for buying assets.

Moving forward

If you have bought a high level of liabilities in the past, try and see if you can use any of them to generate assets. For example, we discussed an avid music fan earlier – he has a huge number of music CDs and an expensive system. It might be possible for him to organise discos for teenagers and earn some cash. Look at what we are suggesting here – we are starting to grow by using what we have to create something new. The argument that you can't move on because you lack a certain item is not valid. You can only start a pattern of development from where you are today, with the tools you have today.

We mentioned Jeffrey Archer earlier – he was effectively swindled out of his money. His response was to look at what he possessed and how he could use that to create wealth. He had a pencil and decided to become a writer and this he has done to great success becoming extremely wealthy in the process. We stress his approach again – use what you have, set a vision, focus on it and then persevere to achieve that vision:

success = vision + focus + persistence

This is what we mean when we wrote about 'paying the price' earlier. The real test of character is to look inside yourself and to decide that you have the focus and the strength of character to persist until you have achieved what you set out to achieve.

> *'If I had to name a driving force in my life, then I plump*
> *for passion every time'*
> Anita Roddick

Unless you are very lucky, don't expected instant success. The Cummutta *Debt to Wealth* programme can take anything from three to seven years. It really does work and is responsible for changing the lives of many people. How serious are you about becoming wealthy? Have you got enough strength of character to persist for seven years, to transform your life from debt to wealth? Only you can answer that.

> *'Man is what he believes'*
> Anton Chekov

Journal	• In your journal, make a commitment to save. We will keep on and on about this because if you fail to take this step you will never become wealthy.
	• Set your target to be wealthy and use the advice in this section of the book and you will be on your way to becoming wealthy.

The next step is to learn how to manage your cash flow forecast.

'Nothing splendid has ever been achieved except by those who dared believe that something inside them was superior to circumstance'

Bruce Barton

Cash flow

A cash flow forecast is a mechanism to predict where your money is going to. This is common management practice in industry – and it is a practice that we can use to manage the household income.

Expenditure	January		February		March	
	Estimate	Actual	Estimate	Actual	Estimate	Actual
Income	£1500	£1500	£1500	£1500	£1500	£1500
Rent	£400	£350				
Food	£200	£195				
Entertainment	£150	£300				
Savings	£150	£0				

Here we have set up a simple cash flow forecast for the first three months of the year. You can set this up in a spreadsheet. There are various ways of portraying cash flow and each has its pros and cons. This version is one of the simplest and we therefore recommend it to you.

In the *expenditure* column you need to list the cost centres – these should match the costs centres in the cash analysis discussed earlier. Each month has two columns – *estimate* for what you expect to happen and *actual* for how things actually worked out. As you work through this month-by-month your estimates should become more accurate.

You need to look at how much you have to spend in total and ensure that you are keeping within your means. Look at the actual spend in January. Savings were estimated to be

£150 but it looks very much as though this went on entertainment. This is *not* how to do it. If you overspend in one area you need to cut in another area but you must try to save.

Friends and enemies

Brace yourself – banks are an enemy. Think about it – the purpose of a bank is to make money, *for the bank*!

Your bank does not exist to make money for you. It exists to make money for its owners and its shareholders. To do this, the bank has a simple job to do – to move money from your account to theirs. They do this by encouraging you to borrow and charging interest on the loan. APR stands for *annual percentage rate*. Disregard what the loan rate states – it is the APR that is important. When you use credit cards or store cards you have an opportunity to pay off an outstanding amount by a certain date or you will have to pay interest. Some store cards carry an APR of around 30%. In other words, if you use a card and do not pay off the cost in the time given, you will be charged almost a third extra for your purchase – so a £30 jumper will actually cost you around £40.

As we said, money is coming out of your pocket and going into the bank's. OK so we may have been a bit strong on this point and banks certainly have a useful role in many ways – but we just wanted to ensure you are aware of the power of the banks to make money and that that money must come from customers.

Compound interest

Let's take a simple example, suppose you invest £100 at 10% for 1 year. At the end of the first year you will have £110. At the end of the second year you will earn interest on the original £100 and also on the £10 interest from the first year. So you are now earning interest on the interest and so it goes on. This can lead to spectacular growth of savings – and debts!

Health warning
Compound interest sounds good – but only for savings. Think about compound interest and its effect on debts – you can end up paying interest on interest. If you are carrying debt on several credit cards it is a good strategy to pay off the minimum on all cards except one. Make this one a target for clearing as soon as possible and then cut it up. The next stage is to do the same with the other cards you have – one, or two at the most, should be able to deal with expenses.

Journal	• You should be working on your cash analysis and cash flow forecast almost daily.

Buying assets

When you have generated a 'pot of loot' you have something to invest. The skill now is to go shopping for assets. In other words when you go shopping, look to buy things that will make you money. In the UK during the late 1990s to 2004 there was a rapid growth in house prices. At the end of 2004 into 2005 prices stopped going up at the same rate, and in fact began to fall in some areas. This has caused nervousness in the housing market because many people bought property as an asset, in other words as an investment. Take professional advice whenever you intend to buy property. Buying property is *not* always a good idea. There are people who have invested in property in the hope that the rising market will make them money – this is dangerous. If the market falls they will fall with it.

In the UK in 2005 there are areas where rents are falling. In London it is now possible to rent an apartment for a lower rent than two or three years ago. There is a danger that the owners of these apartments are actually subsidising their tenants. This is crazy and it is draining them of cash – this is called negative cash flow. As a rule of thumb you should not buy a property to rent out unless you can ensure a rental return of at least 130% of the mortgage price. This then gives you a cushion against increases in the Bank of England base rate. It will also help to finance void periods – those times when you don't have a tenant in your property.

Property is one type of asset

Remember, 'assets feed you, liabilities eat you'. We cannot stress enough that if you are unsure about a potential asset then make sure you get professional advice. Many people who have tried to buy assets have made huge mistakes and it has cost them dear. We do not want you to be one of them. It is simply too risky to buy property anywhere and expect to make money on it, it does not happen that way. Look at the local papers for the area you intend to buy in – are local services good, are there good travel facilities etc. Look at the infrastructure of the town – are there places that would attract the kind of tenant that you can make money from? There are examples of people who have handed over their life savings to companies who have then gone ahead and bought streets of houses in northern England that are effectively worthless. There was a sad case of one man close to retirement who had not even seen the houses he was buying!

In the research for this book we travelled to a particular area of the UK and posed as house purchasers with investment money, simply to ensure that what we were writing was true and based on experience. In this particular area there was an excellent three-bedroom house that was modernised and looked fabulous in the photographs. The agent was keen for us to sign there and then but we insisted on visiting the house. What the agent had not told us was that it was on the edge of a run-down estate where three houses were burnt-out shells. Never buy a house that you have not visited.

We also know of a couple who put their life savings into a 'chocolate box' cottage only to find that it is a money pit. They did not have a survey done on the property and have since found that the roof is collapsing and that the drains need to be dug up and relaid – they flow uphill. This means all of the toilet waste is backtracking to the cottage with all of the unpleasantness that such a situation involves. Professional advice may be costly but it would have saved in the long run.

Pensions – the Mathematics of Taking a Chance

'The past is a foreign country, they do things differently there'
L.P. Hartley

One-minute overview

Pensions are often treated as being unimportant and it is often too late when we realise the need to save. In this chapter we will discuss some basic issues regarding pensions but you *must* take responsibility for your own future. All the advice in this book is designed to make you take action. One of the first things you should do is get professional advice from an independent financial advisor – a small book like this cannot deliver all you need to know. But do not just roll up to the nearest person in your town. Do some homework. When you are choosing a financial advisor, ask for references. If they are not prepared to give you any confidential reports on how good they are and what they have done for other clients, then walk away.

In this chapter we will look at:

- the mathematics of chance;
- annuities and pensions;
- mortgages.

In early 2005, as we were putting the final touches to this book, we met one retired couple in Wales. He is now 79 years old and she is 71 years old. They are retired farmers and still have some land, which is rented out to another farmer. The sad thing here is that they feel trapped. The reason they feel trapped is simple – they invested the money made from selling their farm but they are dependent on keeping the capital intact. Like many retirees, they have a difficult decision to make – if they use up the capital, then they will have less money to live on later in life. This is why many people choose to buy an annuity.

The mathematics of chance

Mathematically speaking, insurance companies work out the probability of how long people will live for. Probability is the mathematics of chance. What is the chance of living to 100? Whatever it is, it is less than the chance of living to 50 years of age. The fact is that more people are alive at 50 than at the age of 100. This means that insurance companies can be pretty sure how long the average man and woman will live. Barring accidents, they have a good idea of the lifespan we can expect and it is on this basis that they can offer different goods at different premiums.

Annuities and pensions

An annuity is a financial product that you buy – in other words you invest your money with an insurance company in buying this product. In return you get a pension for the rest of your life. If you have worked for an organisation and have a pension with that company, you may well be forced to buy an annuity with that money. The advantage of an annuity is that it turns your accumulated capital into a pension and means you will always have income no matter how long you live. The size of that income depends on the nature of the annuity and this is where you will need to seek professional advice.

You need to be aware that if you buy an annuity, the money invested cannot be passed to your heirs on your death. The insurance company guarantees your income for life and the only way to do this is to 'pool' the mortality risk. This means that those who die younger subsidise those who live longer. Some people live for so long that they effectively get 'free' money – get all the invested money back and then some more. But remember, when you die so does your pension.

There are two types of annuity:

- *Pension or compulsory annuities.* When you have paid into a pension plan, on retirement you usually get a lump sum and a ring-fenced amount. The lump sum depends on length of service and is commonly used to clear any outstanding mortgage on the retiree's property. The remaining amount, that we have called ring-fenced, *must* be used to buy an annuity. You have no choice in this– you cannot have this cash.

- *Purchased or voluntary annuities.* This is the type of annuity that anyone with the money can buy.

There are two issues that affect the different types of annuity. Purchased annuities usually give less income than pension annuities. The simple fact is that this is business and in business terms the market for them is less competitive. It is also true that people who consider buying these products tend to live longer. However, with purchased annuities the capital portion of the income is, at the time of writing, tax-free. This can mean that a purchased annuity return, after tax, can give you higher income returns than other types of investment.

It is not possible to give specific advice in this book because there are so many variations – and anyway, you'll be taking some professional advice won't you?. The amount of annuity income that you will receive depends on many factors such as the amount invested, age, sex and the current level of interest rate. The older you are when you make the investment, the higher your income will be. Women, statistically, live longer than men and so get less income. The effective interest rate on the investment is difficult to determine since the insurer will invest it in a number of areas, including government bonds. If interest rates are high then it is usual for the annuity income to be higher, but since the year 2000 interest rates have been low and consequently pensions have not paid out as much as expected.

If you are interested in purchasing an annuity, then seek the advice of a professional and ask for protection against inflation, this is an essential element of any annuity.
Inflation means that your purchasing power is diminished. Take a very simple case of a product costing £1·00 last year and £1·05 this year because inflation across that period was 5%. OK this is hardly an earth shattering increase but there are two issues to consider here:

- your purchasing power is going down – if you have to pay an extra 5p on this item, you will have 5p less for other purchases or savings;

- if you bought 100 such items then you are spending an extra £5.

This is exactly the argument used by the pensioners in the west of England during 2003/4. The local tax in England is called the council tax. This is used to pay for schools, hospitals and local services. In Devon there were increases of 18%. Many pensions are subject to increases due to inflation. At the time that was 2.5% so the 18% tax increase meant that the pensioners' income was falling considerably in real terms. This led to the unique spectacle of older people engaging in civil disobedience in order to make a political point. This was a worry to politicians, in our view, because there is an ever-growing grey vote and the politicians simply cannot afford to lose these potential votes.

Mortgages

The majority of people take out a home loan in order to buy a house. The rate of interest paid on the loan is a percentage figure based on the base interest rates in the market. The base rate is set by the Bank of England. A team of people meet monthly and look at the current interest rate and whether or not it needs to change. Many factors are considered – for example, the Bank wanted to take the 'heat' out of the property market in the UK in 2004. They raised the base rate five times and made people jittery. This slowed the increase in prices and in early 2005 property prices were relatively stable.

Your choice is between a repayment mortgage and an interest-only mortgage, which has some form of investment attached to it which will grow to pay the loan off at the end of the term. From personal experience we do not recommend an endowment policy. There is no longer any tax relief on the life assurance part of the policy and the income return on the funds will have been taxed. If you want to tie your mortgage to an investment it may make more sense to consider an ISA mortgage. There is more flexibility in an ISA than there is in an endowment policy. You will lose far less of your investment by cashing in early – there are no penalties with ISA plans – and, of course, they are free of capital gains tax and income tax. If you already have an endowment mortgage and are considering moving then take financial advice from an independent financial adviser – it may not be a good idea to surrender your endowment policy. The returns will be poor and it is possible to take the endowment policy with you and apply it to part of the second mortgage.

In the early 1990s the standard variable mortgage rate was 15·4% but in the 1950s it had been 4·99%. In 1986 a free market for mortgage rates was possible for the first time. Since the housing boom of the 1980s, the mortgage market has become far more complex with a far wider choice of loan products. If you are considering a first mortgage, a re-mortgage or a new mortgage if you are moving, please be aware that you cannot rely on the lenders themselves to give you the best advice. Generally you can borrow three times the first income plus half of the second income, or two-and-a-half times the joint income – although it is not unknown for couples to borrow up to six times their income. Don't forget that this money has to be paid back, and that if you over extend yourself you may find that you will not be able to afford the repayments should the interest rates go up.

The majority of lenders are prepared to offer you 95% of the property's value and most charge less interest if you have a bigger deposit. Usually you will be expected to have a deposit of between 3 and 10% of the asking price of the property. There are other hidden expenses including solicitor's fees, valuation, arrangement and mortgage indemnity costs. (This is insurance for the *lender* and is payable by *you*. No we have not got this wrong – many lenders insist that you pay insurance in case you cannot pay the mortgage in the future.)

We also advise that if you are intending to buy an old property then the cost of a complete survey is a good investment.

Types of mortgage

The basic mortgage choices are:

- variable rate;
- fixed rate – this is usually higher than the variable rate;
- discount rate – which offers a discount on the variable rate.

Variable rate

The majority of mortgages are subject to a variable rate of interest. This means that when the Bank of England changes the base rate then your mortgage provider alters their interest rate accordingly.

Fixed rate mortgages

These are products that fix your monthly repayment over a set period of time, regardless of what happens to interest rates. At the end of the fixed rate period, your mortgage cost will revert to the lender's standard variable rate. The fact that you know exactly what your payments will be over a period of time means that these products are popular, but use your knowledge of percentages here – are you paying a lot of money for supposed peace of mind?

Discount rate mortgages

These put your interest rate a fixed amount below the variable rate. If the variable rate rises or falls, your mortgage repayments will rise or fall too. This offers some protection from the threat of rising interest rates.

Many fixed rate loans and virtually all discounted offers have a catch. You usually find that you are tied in with your mortgage lender's variable rate for some years after your initial deal finishes. So you are giving up the right to shop around for cheaper deal unless you pay a stiff redemption penalty. These penalties are designed to tie you to the lender after the cut-price period has ended and this is where they claw back the money they spent to get you and make a profit. Remember what we said earlier, *banks are an enemy.*

Your mortgage lender will make you pay a penalty on certain types of loan – usually discounted or fixed interest rate loans – if you want to pay it off before the end of the loan term.

Flexible mortgages

This is a relatively new form of mortgage deal. The flexible mortgage means you can have an account which combines a home loan and a current account. So, if you take out a £150 000 mortgage, and then you win £25 000 on the lottery you can reduce the size of your mortgage without penalty. Flexible mortgages usually come with chequebooks attached. So if you suddenly need an extra £15 000, you'll be able to write a cheque and in the process increase the overall size of your home loan to £165 000. As we keep on saying, take financial advice.

Don't overstretch

The glossy ads are tempting and it's easy to imagine yourself in your new home, but if the interest rate goes up and you default on repayments you will not be in it for long. Lenders do repossess homes and sell them to recover the remaining loan. Don't forget that you can be deceived by interest rates. Taking a loan out when interest rates are low means that you can afford a bigger loan, but if the rate rises you may find yourself in financial trouble.

Paying off the mortgage early

Subject to redemption penalty consideration, it is always a good idea to try to increase your monthly repayment whenever you can. Suppose, for example, you add £100 to your monthly mortgage payment. This makes the loan balance at the end of the month £100 less than it would have been without the extra payment. In the months that follow, you save the interest on that £100 that you otherwise would have paid. The interest payment that you would have made is determined by the interest rate on your mortgage so the yield on your £100 investment is equal to that rate. The rule is that, assuming there is no early payment penalty, principal repayment yields a return equal to the interest rate on the loan. An early payment penalty would reduce that benefit.

Think about this in terms of real life. If you invested that £100 in a savings account you would get a return of something like 3%. If instead you used it to pay a bit extra off the mortgage then your return would be higher – say 5.75% – because mortgage interest rates are always higher than the base rate. Also, bank savings accounts are subject to tax so the difference between saving the money and paying off the mortgage early is even greater

'The end justifies the means'
Hermann Busenbaum

One-minute overview

At the end of a book like this it is easy to finish and ask 'What have I learnt?' So we thought it would be useful to recap some main ideas.

In this chapter we will look at:

■ developing your mindset;
■ planning for your future;
■ debt to wealth;
■ ordering priorities;
■ principles and values;
■ your future.

Developing your mindset

Your mindset is the way you think. Do you live in a world of scarcity? Is there never enough money to go around? Or do you live in a world of abundance where there is more than enough for everyone?

It is a fact that if you have the wrong mindset you will never be successful. Steven Covey is an expert in relationships and wrote *The Seven Habits of Highly Successful People* – an international best-seller. Covey recommends that you base your practice of day-to-day living on principles and values. He advocates, as one of the habits, a 'win–win' situation. Whenever you are in negotiation with another person over a matter, try to ensure there is a successful outcome for both parties.

This is how the right mindset is put into practice. When you have the mindset that the world is a place of abundance with plenty of opportunities for all, you do not need to bleed every last benefit from a relationship. By thinking 'win–win' you are ensuring that you achieve your objective, and also that the person with whom you are negotiating benefits. This is a mindset based on values and principles. One of the values here is that everybody deserves respect.

Covey's Habits

Covey's first habit of highly successful people is to be proactive. Proactive people are driven by values and use their resourcefulness to be troubleshooters. They solve problems rather than waiting for other people to solve them. There's a good example of this is the company we use to typeset our books. Marian, one of the typesetters, recently phoned the Studymates office with a problem regarding one of the books. She immediately suggested a solution to the problem and saved the editorial team a great deal of time. All the editorial team had to do was agree with what she suggested. Marian is a proactive professional. Proactive people focus their efforts on where they can be most effective. This can often increase their sphere of influence.

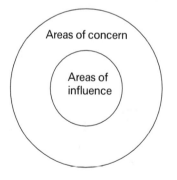

The diagram shows the areas of concern in our lives, whereas the smaller circle is the area of influence that we have. Negative, or reactive, people often focus their efforts in areas over which they have little control. They waste energy and emotion in complaining and, effectively, their area of influence shrinks. As Gerry Robinson writes in his book *I'll Show Them Who's Boss*, 'Don't sweat the small stuff, concentrate on where you can be most effective'.

Here are Covey's Seven Habits.

1 *Be proactive.* We can only change ourselves – change comes from within. Highly effective people change their lives and improve their lives by influencing things they can influence, rather than constantly being reactive to outside forces.

2 *Start with the end in mind.* In terms of this book, the end we want you to have in mind is being financially debt-free and ready to start investing. Develop your own mission statement – for instance, 'Our purpose is to create a debt-free family with values that are inclusive for everyone.'

3 *First things first.* It is very easy to be busy doing the wrong things – psychologists call them 'displacement activities'. The writer Iris Gower tells a story of how, when she was writing her early novels, she would take the children out of the playpen and climb in herself with her typewriter. The children roamed freely around the room while she stayed behind the bars writing. Don't manage time by trying to fit more and more into your day. The fact is that you are limited by 24/7, like everyone else. Strike a balance between work and family, identify the key roles you take on and make time for each of them – this is what living by values and principles means.

4 *Think win–win.* Seek agreements that are mutually beneficial – we think this is the most important habit in developing productive relationships that are more likely to achieve progress.

5   *First seek to understand, then to be understood.* Your first aim should be to understand the other person's point of view. Listen intently to what they are saying and try to put yourself in their position. The US writer Jay Abraham tells a story where he met a man in Australia. The only things he said to this man was his name and where he was from. At the end of their conversation the man told Jay how interesting *he* had been. The fact is that people do like to talk about themselves and, as long as you use this habit ethically, it can lead to deeper and more fulfilling relationships both personally and professionally.

6   *Synergise.* By leverage through relationships that are based on mutual trust, it is possible to create something that is greater than the sum of the parts. It is also easier to solve conflicts and find a better solution.

7   *Sharpen the saw.* Everybody occasionally needs something in life that renews the spirit. There is one student who has a unique way of renewing her spirit – she was the life model for the local art class. For her this was the ultimate way to both relax and have her self-esteem boosted. While this may not appeal to you, there must be balance in your life where you have some guarded time to do whatever it is you like doing.

In late 2004, Covey announced an eighth habit – 'give people their voice'. This means allowing people, with whom you work, to take control and responsibility for their area. If this is done sensitively and with care, it will leverage performance and make people feel more valued.

Dr Covey has done us a great service in isolating these habits and we recommend his books to you – they are based on principles and values rather than personality..

Yelling at people

I have just read a book by a well-known American businessman who advocates that it is necessary to yell at people from time to time. This is something many people have great difficulty accepting. If you are yelling at people then you have failed to make your case and are, frankly, resorting to bully tactics.

> 'I have a visceral reaction to bullies, I can't stand it when a
> predator takes an unfair advantage'
> Rudolph W Giuliani, Former Mayor of New York

So in developing the right mindset for your success, think of other people and how they too would benefit. Think interdependence rather than independence. Dr Stephen Covey's work has shown us that, rather than have personality-based ethics like the US businessman who advocates yelling at people, if we develop character-based ethics we will achieve more and have happier and more fulfilled relationships and lives.

Planning your future

Here is your first opportunity to use Covey's principles. Start with the end in mind (*Habit 2*). What do you want to achieve? We encouraged you to start a journal to record your aspirations and progress. Not done it yet? You need to now.

There is a famous case of a class of 1950s' American graduates. They had a reunion about 25 years later and it was found that only 3% of them had written down targets. This 3% had planned their future by writing down what they wanted to achieve and kept their journal active throughout the years. The remaining 97% had not done so. When they added up the worth of the 3% they found them to be worth more than the combined assets of the remaining 97%. This is simply because they had an end in mind and had been proactive rather than reactive (*Habit 1*).

Planning your future is like sailing a yacht across a bay – depending on the wind direction, you would probably need to tack from side to side. The same is true of your future. You need to be flexible. Plans you currently have will probably be altered, or even dropped, as you, your understanding and your education develop. This is normal – the only constant in life is change. Just make sure that you are proactive in this respect. One way of doing that is to commit and to write your plans, and their changes, down. In this way they become real tangible goals that you can *see* and from which you can gain inspiration.

If you want to see this in action, watch the movie *Educating Rita* by Willy Russell. We see Rita develop from a very insecure person to one who is confident and competent in what she is doing. She takes control of her future, drops the name Rita and reverts to her real name. Julie Walters plays Rita brilliantly and does us all a service in showing how education can change someone's life.

Improve your education

This does not mean going back to school or taking exams – it does mean *learning*. We know of one businessman who was considering a ludicrous venture – or so it seemed to us. It was suggested that he should consider participating in a training course, or at least buy some books. He scoffed at these ideas saying that he had 30 years' experience. He went ahead with the venture and lost £250 000. The point is that had he invested some time to collect advice or £50 in some relevant books, he would have seen that his plan was flawed.

The fact is that the world *has* changed. As we noted that at the start of this book, we have moved from the industrial age to the information age. The fact that this business man had 30 years experience was irrelevant – he did not have 30 years of pertinent experience. No one does. Don't listen to the pub expert who has never invested or planned in his life

– educate yourself and learn to think for yourself. That is what we mean by education.

Debt to wealth

It is important to identify areas of expenditure that you can cut out. In particular, look at areas where you are buying liabilities rather than assets. Use the spare capacity to pay off debt and create a pot of cash to move on and invest.

Order of priorities

Your first priorities must be debt repayment and protection. You need to ensure that your dependents are protected. Make sure you have enough life assurance, take advice from an independent financial advisor and shop around – there are bargains to be had.

- If you are unsure what to do about debt repayment then visit your Citizen's Advice Bureau. They have trained debt counsellors who can advise you accordingly, and their services are free. You need to improve your cash flow management and keep your spending below your income. This will create spare capacity to pay off your debt – and then to create cash to invest.

- If you have no dependents then insurance and assurance are next in line. But don't wait until your debts are repaid to get insurance or assurance if you have dependents. The risk is too great. 'Assurance' is the term used to relate to something that will eventually happen. The classic is life assurance – we will all die one day.

- Create an emergency fund. You should ensure that you have enough money in it to live on for at least three months. This will give you a safety net to get back on your feet if an unexpected event knocks you off track – and they do happen.

- Start investing but be cautious. Get some education – make sure that you understand where you are going and why. We have mentioned a number of cases in this book where people have been ripped off. Don't abdicate responsibility for your own future, make sure you get good quality advice and follow it.

Principles and values

Decide which principles and values you will use as the bedrock of your plans. As an example of this, here is an extract from the *OnTheIssues* website (www.issues2000.org).

> TheThirdWay philosophy seeks to adapt enduring progressive values to the new challenges of the information age. It rests on three cornerstones:
>
> ■ the idea that government should promote equal opportunity for all while granting special privilege for none;
> ■ an ethic of mutual responsibility that rejects equally the politics of entitlement and the politics of social abandonment;
> ■ a new approach to governing that empowers citizens to act for themselves.

This is a set of values that guide the New Labour philosophy in the UK and the Democrat philosophy in the US. What are your principles and values?

At Studymates, we have a principle that a reader is entitled to good writing and to have information presented clearly. Student books, in particular, have to be checked rigorously for accuracy. Experts examine the work and make sure it is accurate.

Your future

The future is exciting if you are prepared. At the start of

this book we said we have to face up to the events expected in 2010. It will have serious economic repercussions around the world.

You can never know too much. Think of thoughts as tennis balls with hooks on them. They link thoughts. The more tennis balls (or thoughts) you have the more hooks you have and the more connections you can make. We know of one woman in the south of England who worked on a supermarket checkout. Now there is nothing wrong with jobs like this and those who do them provide a valuable service. But she was unhappy with the fact that there was never enough money in the house. In her spare time she trained as a driving instructor and started a new part-time career. She now, less than five years later, runs a highly successful driving school with more than ten cars and instructors working for her.

Taking action

By 'taking action' we mean making a start to changing your life – but, like the budding driving instructor, don't give up the day job! In the early stages you will need the security of that source of income.

One storeman in Leicestershire has done exactly that. He has:

- gone to night school for a number of years to learn about investing;

- invested in companies *only after doing his homework* and making his own decisions;

- made over £40 000 between 2000 and 2005, by carefully investing and reinvesting;

- kept his day job.
He is an ordinary man who left school at 16 but he got himself a great education and is a major success, but he still has the day job. His story is inspirational.

The bottom line

Remember:

**ideas + beliefs + commitment + action
= enhanced personal development and self-growth**

You are where you are in life because of the decisions and choices you have made so far. If you keep on doing what you have always done, you will keep on getting what you have always got. The only way to change your life is to take control.

We hope we have given you the inspiration to take the first few steps.

Fast Fax

1   1 inch is about 2.5 cm; 1 foot is about 30 cm.

2   1 kilogram is roughly 2 lb. For a more precise conversion from kilograms to pounds, first divide by 2 and then subtract 10%.

3   1 euro = 100 cents.

4   Area is a measure of 2-dimensional space.

5   Volume is a measure of 3-dimensional space.

6   To save 10% of your income, divide the total income by 10 (then put this amount into a high interest account and leave it).

7   Quick VAT calculations (currently 17.5% in the UK): For example, suppose you want to calculate the VAT to be added to £10·00

   first divide by 10 – this gives £1·00
   now divide that by 2 – this gives 50 p
   now divide that by 2 – this gives 25 p
   add all three amounts together – this gives £1 + 50 p + 25 p = £1.75

   So the VAT on this £10·00 item is £1·75, and it costs £11·75.

Index

JAN 2 4 2007